Digestion

Digestion

Fueling the System

By Dr. Gordon Jackson
and Dr. Philip Whitfield

TORSTAR BOOKS
New York • Toronto

TORSTAR BOOKS INC.
300 E. 42nd Street
New York, NY 10017

THE HUMAN BODY
Digestion:
Fueling the System

Publisher
Bruce Marshall

Art Director
John Bigg

Editor
Beryl Leitch

Art Editor
Peter Bridgewater

Text Editor
Gwen Rigby

Researcher: Jazz Wilson

Director of Picture Research
Zilda Tandy

Picture Researcher: Jessica Johnson

Artists: Frank Kennard, Mike Courtney,
John Bavosi, Andrew Popkiewicz,
David Gifford, Les Smith, Ivan Hissey

Art Assistant: Jean Foley

Cover Design: Moonink
Communications

Cover Art: Paul Giovanopoulos

Director of Production: Barry Baker

Production Coordinator: Janice Storr

Business Director: Candy Lee

Planning Assistant: Avril Essery

International Sales: Barbara Anderson

In conjunction with this series Torstar Books offers an electronic digital thermometer which provides accurate body temperature readings in large liquid crystal numbers within 60 seconds.

For more information write to:
Torstar Books Inc.
300 E. 42nd Street
New York, NY 10017

Authors

Gordon Jackson is a consultant in Internal Medicine at Lewisham Hospital, part of the Guy's Hospital Medical School. His early research was cardiological in nature, but his recent work has been concerned with intestinal absorption and allergy. He has also written extensively on health related topics.

Philip Whitfield is a biologist and zoologist with a wide range of interests. He has written and contributed to many books, including *The Animal Family, Jungles, Rhythms of Life* and *The Biology of Parasitism.* He has also published papers in academic journals.

Chris Allen is Senior Registrar in Neurology at the Charing Cross Hospital and Medical School in London. He came to neurology after considerable experience in general internal medicine and has carried out research on many neurological topics, especially stroke.

Mike Lavelle is a Consultant Surgeon at Cuckfield Hospital, Sussex, England, specializing in abdominal and urological surgery.

Liz MacFarlane specializes in family medicine, particularly the problems of adolescence. Her current research is in the field of hypertension. She has written on a wide range of health related topics.

John Rees is a clinical tutor in the United Hospital Medical Schools (Guy's Hospital campus). He is a specialist in chest medicine and internal medicine and has published extensively in the field of asthma.

Mike Stoddart is a biologist and zoologist with particular interests in mammalian behavior and the study of olfaction. He is the author of *The Ecology of Vertebrate Olfaction* and has also contributed to many other books on the natural sciences including *The Animal Family, Jungles* and *Rhythms of Life.*

Consultants for Digestion

Jerry Nagler is currently Clinical Assistant Professor of Medicine at Cornell University Medical College. He is also Assistant Attending Physician at the New York Hospital, a Fellow of the American College of Gastroenterology, a Member of the American Gastroenterological Association, a Member of the American Society for Gastrointestinal Endoscopy and Cochairman of the New York City scientific advisory committee of the National Foundation for Ileitis and Colitis.

Michael J. Schmerin is Clinical Instructor of Medicine at Cornell Medical College and on the attending staff at New York Hospital and Lenox Hill Hospital. He has published many articles in the field of gastroenterology.

Charles S. Winans is Professor of Medicine and Codirector of the Section of Gastroenterology at the University of Chicago Department of Medicine. He has written widely on the subject of gastroenterology in many books and journals.

Series Consultants

Donald M. Engelman is Professor of Molecular Biophysics and Biochemistry and Professor of Biology at Yale. He has pioneered new methods for understanding cell membranes and ribosomes, and has also worked on the problem of atherosclerosis. He has published widely in professional and lay journals and lectured at many universities and international conferences.

Stanley Joel Reiser is Professor of Humanities and Technology in Health Care at the University of Texas Health Science Center in Houston. He is the author of *Medicine and the Reign of Technology;* coeditor of *Ethics in Medicine: Historical Perspectives and Contemporary Concerns;* and coeditor of the anthology *The Machine at the Bedside.*

Harold C. Slavkin, Professor of Biochemistry at the University of Southern California, directs the Graduate Program in Craniofacial Biology and also serves as Chief of the Laboratory for Developmental Biology in the University's Gerontology Center.

Lewis Thomas is Chancellor of the Memorial Sloan-Kettering Cancer Center in New York City and University Professor at the State University of New York, Stony Brook. A member of the National Academy of Sciences, Dr. Thomas has served on advisory councils of the National Institutes of Health.

**Library of Congress
Cataloging in Publication Data**

Jackson, P.G.
  Digestion

  Includes index.
  1. Digestion. 2. Digestive organs. I. Whitfield, Philip. II. Title.
QP145.J33 1984 612'.3 84-15765
ISBN 0-920269-38-9

ISBN 0-920269-22-2 (The Human Body series)
ISBN 0-920269-38-9 (Digestion)
ISBN 0-920269-39-7 (leatherbound)
ISBN 0-920269-40-0 (school ed.)

20 19 18 17 16 15 14 13 12 11
10 9 8 7 6 5 4 3 2 1

Contents

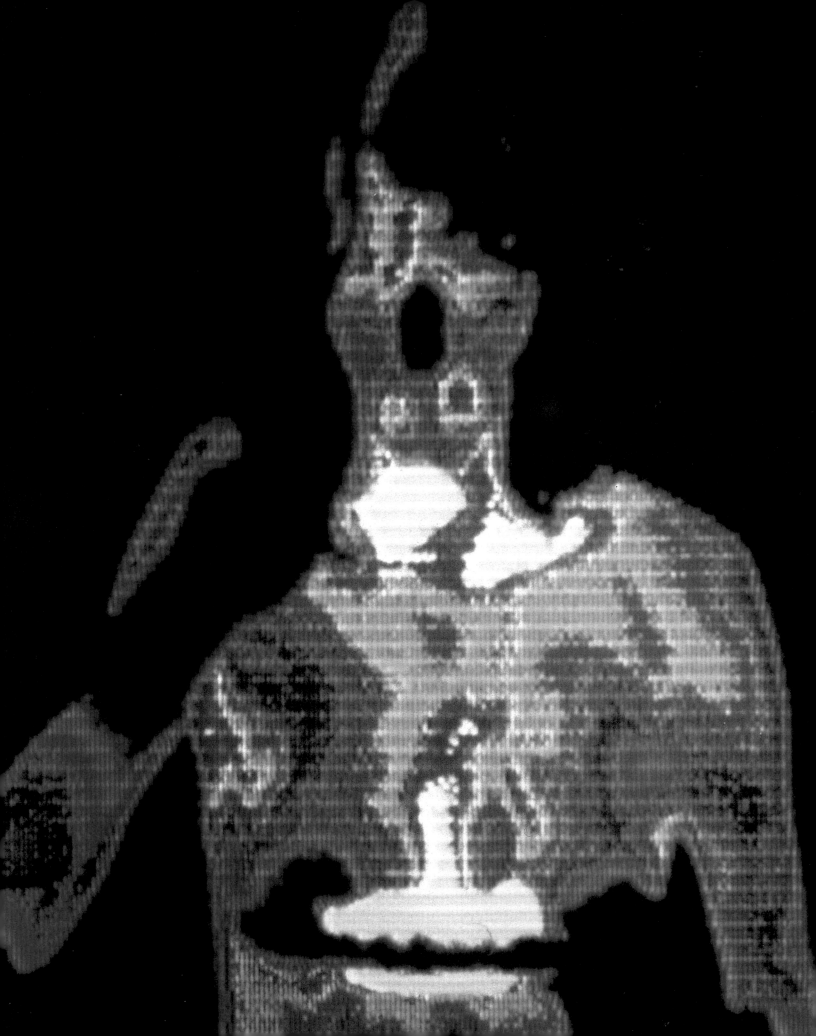

Introduction:

The Substance of Life

Eating is one of life's great pleasures. Yet food is more than a sensual delight, it is essential to human survival. Without the urgent spur of appetite, and deprived of a regular supply of food, life's candle soon flickers and is snuffed out.

Food is needed to provide the body with all the nutrients it needs; with chemicals for body maintenance and repair; and with the essential raw materials for fighting off infection. But if food is to be of any intrinsic value to the body, it must first be digested, and a digestive system that is working at its maximum efficiency is vital to physical well-being. For as Shakespeare knew well:

> Now good digestion wait on appetite
> And health on both!

In the digestive system, which in essence is no more than a long tube traversing the interior of the body and flanked with glands that pump in their products to aid its work, food is ground, churned and liquefied. Food is mixed with a staggering laboratory of chemicals, which break it down in a series of complex reactions into small molecular units. These can be absorbed into the blood and swept away in the current of the circulation to the parts of the body where they can carry out their appointed tasks. At the same time, useless and indigestible portions must be prepared ready for elimination from the body.

A healthy digestive system is one of which its owner is almost unaware. Only when things go wrong do pain and indigestion send out warning signals that all is not well. In the past, the very invisibility of the digestive system was an enormous obstacle to the diagnostician, but today technology makes it possible for the gastroenterologist to view the inside of the gut with remarkable clarity. Surgery, too, has made such dramatic advances in safety and technique that conditions once feared as killers can today be treated as routine, giving sufferers unexpected, bonus years of high quality living.

Thermographic scanning, one of the many technological advances in the world of medicine, allows specialists to inspect the digestive system in action. Such methods give valuable insights into the journey food must make as it is transformed into the molecules that fuel existence. The following pages are devoted to charting and explaining that journey.

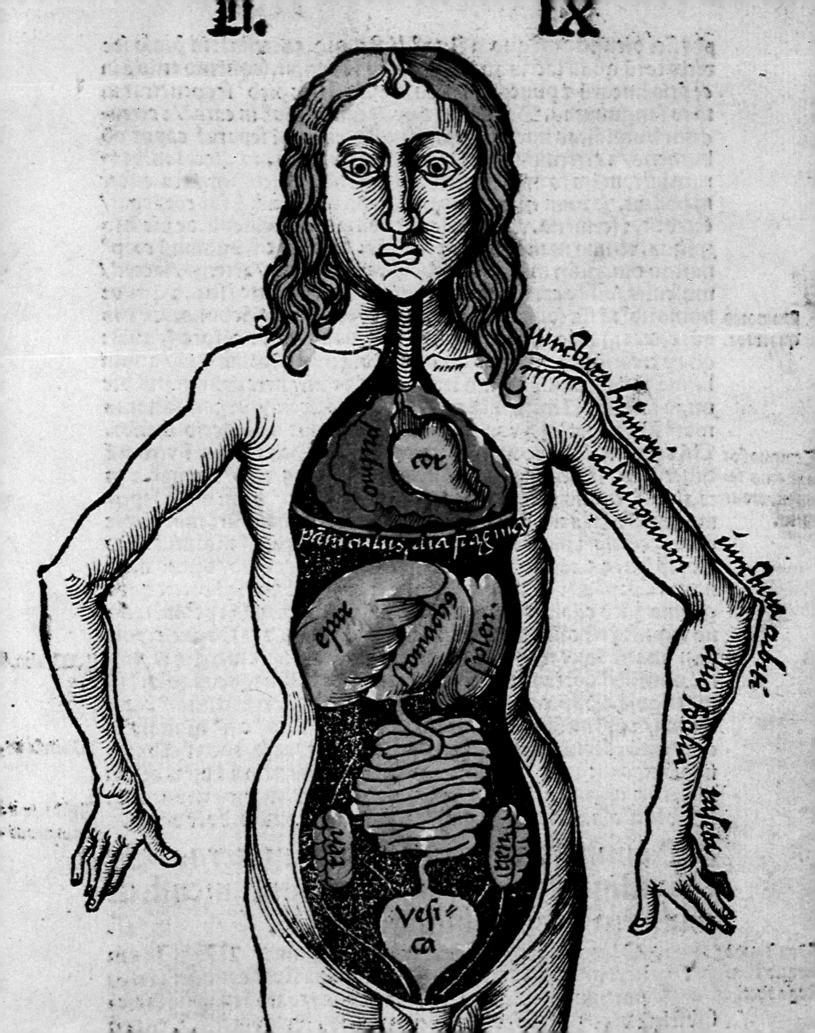

Chapter 1

The Pathway of Alchemy

A green plant, using chlorophyll and sunlight, can synthesize living parts—leaves, roots, flowers — from water, carbon dioxide in the air and mineral salts in the ground. The organic substances thus produced are, essentially, the base of the entire food chain, for animals are unable to create organic substances from an inorganic starting point. Instead they must make use of the plant created substances either at first hand, by eating them, or at second or even third hand, by consuming other creatures that have in turn eaten plants.

With this in mind, it is to be expected that the absorptive mechanisms of animals differ dramatically. Indeed, they range from the relative complexities of the omnivorous human being to creatures that do not even possess any differentiated internal structures dealing with digestion: in fact, many organisms do not possess a gut, the part of the body concerned with digestion.

Surviving Without a Gut

The single flagellated cells of the trypanosome protozoan, *Trypanosoma*, which cause human sleeping sickness, swim constantly in the blood of the sufferer. These single celled parasites live in a carefully controlled superabundance of simple nutrients like glucose and amino acids within the blood plasma. The *Trypanosoma* can remove a nutrient such as glucose from the blood at an incredible rate — the parasitic cells can absorb their own dry weight in glucose in between one and two hours — but the sufferer's own metabolic control mechanisms carefully and rapidly replace these lost nutrients, providing additional nourishment for their unwanted guests in the process.

Parasitic lifestyles appear to have been associated with many of the most extraordinary examples of gutless animals. This is almost certainly because so many of the host's body tissues and fluids provide easily absorbed food. For *Trypanosoma*, the

This 1508 illustration, complete with the inaccuracies due to lack of knowledge, shows by its mistakes and omissions just how far we have come toward understanding the life-sustaining digestive system.

9

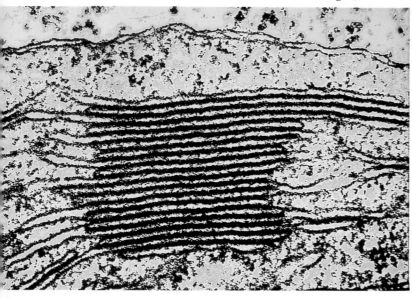

The chloroplast, an organelle within the cells of plants, contains chlorophyll, a green pigment which traps the energy of sunlight and enables the plant to synthesize living molecules (below).

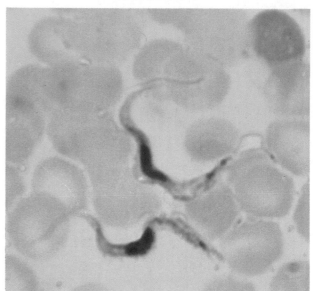

Trypanosoma rhodesiense, *the parasite responsible for the condition of sleeping sickness in humans, swims about in its own nutrient medium — human blood (below right).*

Worms such as those of the group Pogonophora have dispensed with the need for a gut. They live by warm vents deep on the sea bed and absorb food directly across their body walls (bottom).

nutritious fluid is blood, but there are many other mediums. Several species of adult tapeworm, for instance, live in the small intestine of vertebrate host animals. Although these white, squirming parasites may reach several yards in length, they do not possess a gut. All nutrients are taken in through an elaborately modified body wall which contains millions of microscopic fingerlike projections — modified microvilli called microtriches — which increase the absorptive area.

Another group of parasitic worms, the spiny-headed worms, or acanthocephalans, lives in the same area as the tapeworms. And although un-related to the tapeworm, acanthocephalans have also lost all trace of a gut. To increase their surface area for food intake, their bodies are indented with myriads of branched pore canals which turn their outer covering into something resembling a sponge that soaks up nutrients.

Not only parasitic or one celled animals manage to exist successfully without a digestive tract. Clustered around natural hot water vents deep in the Pacific Ocean are the pogonophorans — a bizarre name for bizarre animals. These extraordinary, deep sea worms are believed to be distant evolutionary relatives of the commonplace earthworm. Over a yard long, they have no gut at all, absorbing nutrients instead across their body wall.

For most animals, however, the provision of

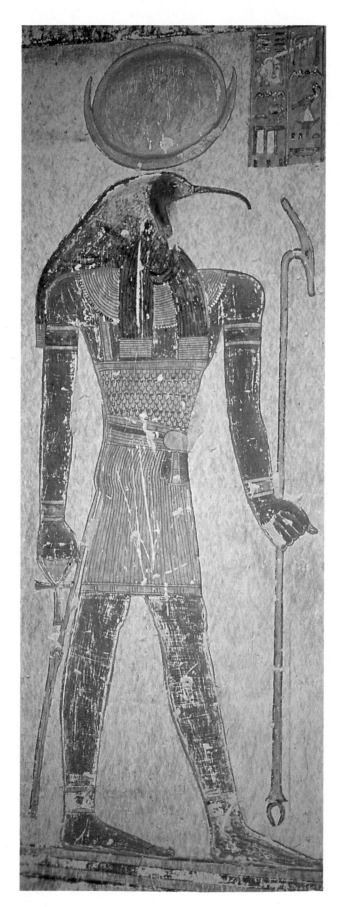

Physicians in ancient Egypt looked not only to their supposedly curative concoctions but also sought guidance from the god Thoth, depicted left in a wall painting from the tomb of Rameses VI.

nutrients for growth, bodily repair or energy supply is a more complex matter. Most animal foods come in the form of large molecules, such as those of starches, proteins and fats, which are too large to be easily absorbed. During the process of digestion, these large molecules are broken down into their ultimate organic building blocks which can more easily be assimilated. Food passing along the path of the human alimentary tract undergoes a spectrum of changes, induced by mechanisms of which, in the main, we have no conscious experience, but whose coordination is a prerequisite for our continued existence.

The anatomical passage from mouth to anus can be described by that short and direct Anglo-Saxon term — gut. Digestive tract, alimentary canal and digestive system seem more scientific descriptions, but gut is a simple and concise and correct word for this route of food transformation.

Knowledge of the human digestive system and its operation has a long and complex history. We can chart here only some of the most conspicuous of the recorded phases in the history of the accumulation and refinement of medical and scientific information about the gut.

The History of the Gut

As in so many other areas of historical investigation, for the first documented evidence of knowledge about the human alimentary tract we must return to Ancient Egypt and to the more or less contemporaneous civilizations in the valleys of the Tigris and Euphrates — Mesopotamia — the land between the rivers. Among the written records of the period, both on paper and inscribed stones, it can be found that there was already considerable knowledge of the gut and how it worked.

A varied range of medicines was available to the physicians of Egypt operating under the guidance of the god Thoth, who was said to give physicians skill to cure. The drugs were usually made from herbal, vegetable or animal products, but salts of antimony and copper were also used.

One of the best preserved medical papyri is the Ebers Papyrus, written in about 1570 B.C. It is comprised of one hundred and ten pages, lists nine hundred prescriptions and was found in a temple at Thebes, in 1862, by Professor Georg Ebers. It is

the oldest complete medical book in existence and has been claimed by some authorities to be the oldest entire book of any type still to exist. Among other things it contains many suggested treatments for stomach and intestinal worms, such as hookworms and *Ascaris*, as well as for the gut symptoms caused by *Schistosoma mansoni*, a blood fluke that inhabits the mesenteric blood vessels close to the small intestine. Egyptian physicians of this period had already developed the now familiar technique of introducing drugs into the bowel in the form of suppositories for illnesses of the lower bowel.

The wedge shaped, cuneiform writing on clay tablets and stone of the Mesopotamian civilizations also gives us a fascinating insight into the alimentary medicine of early times. Anatomical knowledge of the liver was extremely well developed in this culture over four thousand years ago, primarily because the livers of animals such as sheep were

important to the diviners, who understood omens and used the liver for their predictions.

During the periods of the ancient Greek medical traditions of Aesculapius, Hippocrates, Aristotle and Galen of Pergamum (A.D. 129–199) work on the gut continued. Galen's conceptual framework was to dominate medical thought and practice for thirteen centuries after his death. Galenism, the name given to his teachings, is not easy to define. However, we can gain an insight into his view of the physiology of the gut.

Galen thought that the life principle was a spirit, or pneuma, drawn in while breathing. It entered the pulmonary veins and from there the left side of the heart where it was supposed to meet the blood. In the area of digestive physiology, Galen postulated that food material from the gut was transported by the hepatic portal vein to the liver and was there converted to blood and linked with its own version of the vital pneuma. Among other problems in this supposed design of things was Galen's lack of knowledge of the circulation of the blood. In his opinion, all that the muscular pumping of the heart did was to produce a tidal motion in the blood to mix it with pneuma.

In the seventeenth century, Galenism was under serious attack as a framework of knowledge and understanding, and, by the end of the eighteenth century, it no longer had standing as an intellectual or academic force. In this dynamic period of change, a considerable amount of important research with specific significance for the understanding of the human digestive system was undertaken. In the last ten years of his life, the French naturalist René de Réaumur (1683–1757), well known for his invention of an alcohol thermometer and establishment of a temperature scale, made many important discoveries about the gastric juices of birds. He showed that juices from the stomach, when held at body temperature outside the body, have the power to dissolve food. Lazzaro Spallanzani (1729–1799), in Italy, showed that digestion was a different process from either fermentation or putrefaction and that the stomach itself could secrete digestive juices.

It was the English doctor William Prout who showed, in 1824, that hydrochloric acid was also present in the stomach; while the American army

William Beaumont

The Backwoods Physician

"Facts are more persuasive than arguments," so stated the U.S. Army surgeon William Beaumont (1785–1853), reporting his observations on the secretion of gastric juice in 1833. As often occurs in the history of science, it was chance that gave this careful observer his material for study. Beaumont was called in June 1822 to see a French Canadian, Alexis St. Martin, who had been seriously wounded in a shotgun accident. Although apparently mortally wounded by a shot in the stomach, St. Martin eventually recovered but was left with a permanent hole, or fistula, from his stomach to the skin of his abdomen. Through this opening, the lining of the stomach could be seen, and the changes in secretion that occurred under various stimuli recorded.

When the young man was recovered from his wound, though still with the fistula, Beaumont was able to start his series of crucial experiments.

Before the direct observations made with such care and determination by Beaumont, much argument wracked the medical world as to the nature of the gastric juice and the control of its secretion. Beaumont was able to isolate the acid secretion of the stomach and to demonstrate

that it was produced not constantly, as had been maintained, but in synchrony with the presence or anticipation of food in the stomach. Beaumont confirmed that the acid made by the stomach was hydrochloric acid and that the acid secretion was produced separately from the mucous secretions of the stomach.

He observed and recorded the effects of mental changes on the production of gastric juice and the movements of the stomach. He also showed the adverse effects of certain foodstuffs on the stomach's lining, in particular the inflammatory consequences of alcohol abuse. He showed that, in addition to the acid that the stomach made and secreted,

another substance active in digestion was also present. Two years after Beaumont's publication of his findings, Theodor Schwann showed this to be the enzyme pepsin.

It was eventually recognized that Beaumont had made considerable advances in the understanding of digestion. Indeed, his work pioneered the study of the physiology of digestion that was subsequently carried out in the second half of the nineteenth century.

William Beaumont has since been much honoured, and the famous physician Sir William Osler described his work as "a model of patient, persevering investigation, experiment and research." Others have said that "Every physician who prescribes for digestive disorders and every patient who benefits by such a prescription owes gratitude to the memory of William Beaumont."

Beaumont continued his hard work as a surgeon right up to the weeks just before his death in 1853. By this time, his book on the physiology of gastric secretion was a classical work, translated into many languages, and, by the end of the century, was the basis for the descriptions of digestive physiology found in most medical textbooks.

Dr William Beaumont (below),
was able to observe the interior of the
stomach of the unfortunate Alexis St
Martin, left with a fistula (hole)
between stomach and skin after a
shooting accident.

Life for filter feeders — creatures who
sit tight waiting for waterborne
nutrients to come their way — is
undemanding. The sea squirt, Ciona
intestinalis (bottom) is a good
example.

surgeon William Beaumont (1785–1853) had the opportunity to perform experiments in gastric physiology which laid the foundations of understanding of the function of the stomach in digestion.

To place this remarkable organ complex in its proper context before examining in detail its structure and function, two areas need to be examined, both related to types of development. The first is the comparative backdrop provided by the human gut. In other words, how does our digestive tract fit into the inventory of other mammalian guts? The second is the context of an ontogenetic framework — how does the digestive system that we see in a child or an adult arise during the developmental wonders in the womb after fertilization?

Gut and the Diet

Zoologically speaking, humans are omnivores — we feed on a mixed diet of animal and plant food. Accordingly, we have a mixed dentition consisting of incisor teeth, canines, premolars and molars. Our teeth are very different from those of a dog, which is a carnivore or meat eater, or those of a cow, which is a herbivore or plant eater. Because we are omnivores, with an unspecialized diet, our own digestive system serves as an example of an unspecialized gut. Many — indeed most — of our mammalian relatives are carnivores or herbivores, enjoying a more specialized diet than we do and showing marked adaptations because of this.

Among lower animals — worms or insects — there are examples of filter feeders and fluid feeders, but these are virtually absent from the mammals. Filter feeders are equipped with a sievelike structure which is wafted or dragged through water and strains out microscopic animals and plants. The only filter-feeding mammals are the baleen, or whalebone, whales.

As anyone who strays outside on a warm summer night knows, many insects feed by sucking blood and even more feed by sucking plant juices. Among the mammals, only the vampire bat has adopted this habit of sucking blood, but substantial modifications have occurred to its digestive system, enabling it to thrive on this restricted and specialized diet. Some humans exist for long periods of time on a fluid diet, notably the Masai tribesmen from Kenya who feed on blood and milk from their cattle from time to time, but eventually even they require roughage and bulk.

There seems little doubt that man's ancestors were much more carnivorous than we are today. They supplemented their largely meat diet with seasonal fruits, berries and shoots. However, the gut and digestive system of modern man is less specialized than that of his ancestors and better adapted for a general diet.

The Carnivores

The food of carnivores is nutritionally extremely rich. Animal flesh is essentially protein and, with the exception of inedible bones, hair and feathers, completely digestible. Carnivorous birds regurgitate the bones and hair in a cylindrical pellet soon after feeding so that no unnecessary weight is carried into the air. Carnivorous mammals simply eat everything, and their feces contain substantial and identifiable amounts of inedible fragments. Hyenas chew up the bones of their victims, rendering them into a white paste which gives the feces a distinctive

appearance. This bonewhite feces, when ground up with a little oil, produced the white pigment Album Graecum used by the Ancient Greeks to make the tablets upon which they wrote.

In contrast to herbivores, such as the cow or horse, which literally have their food constantly at their feet, carnivores have to expend much energy to acquire food, and it is available only intermittently. For this reason the gut, and particularly the stomach, is designed to be empty one moment and overgorged the next. For instance, a dog's stomach is simple in structure and highly distendable. And in some species, such as the Cape hunting dog from southern Africa, the stomach is used as a store cabinet. Meat is carried from the hunting ground to the breeding ground in the dog's stomach and then regurgitated in huge chunks, which are instantly snapped up by the pups and their minders.

Meat is easy to assimilate. Proteolytic enzymes rapidly reduce the meat in the stomach to a liquid chyme which is then released into the small intestine for digestion. The gut of carnivores is fairly

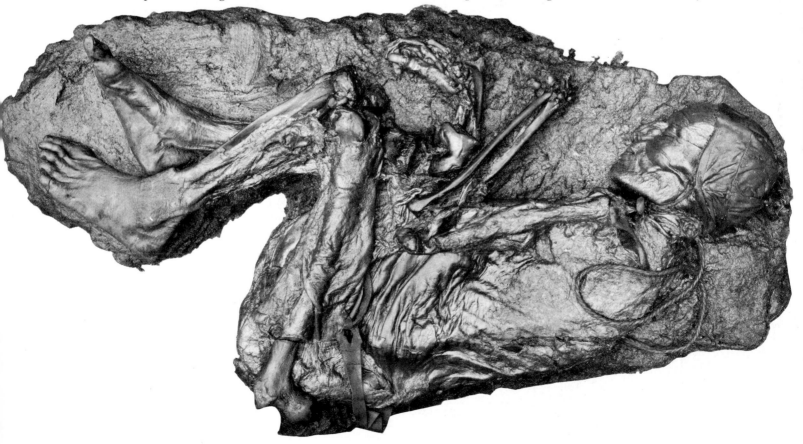

short, for absorption occurs rapidly. This is in marked constrast to the gut of herbivores. Plant material is difficult to digest and therefore requires an extremely long gut.

There are few carnivores which have changed in the course of time to become herbivores, although there are many examples of herbivores adopting a more or less carnivorous way of life. By far the best known example of carnivore-cum-herbivore is the giant panda. The exact zoological origins of this animal are still a matter for discussion, but it is clearly related to the raccoon, the lesser panda and the bear. These three groups are strongly, but not exclusively, herbivorous. Pandas, in fact, feed on a restricted and specialized diet of bamboo shoots and leaves. When Chi-Chi, the giant panda at London Zoo, required abdominal surgery, vets

were amazed to note how short the gut was for a herbivorous species. The rarity of this attractive species is perhaps because it has become over-specialized and cannot now adapt to changing ecological conditions.

The huge baleen whales of the world's oceans are highly specialized carnivores. Zoologists are unsure about the origin of this group of mammals. Recognizable whales existed in the Eocene epoch, long before many of the mammals we see today were around, and we know that they evolved from land mammals, who doubtless took to the sea in search of crustacean food and never returned.

Whales are interesting because, while they are carnivores and feed exclusively on animal tissue, they are also filter feeders. Their food is krill — a small shrimp up to two inches in length —

which abound in huge swarms in the nutrient rich, but cool, waters of the Antarctic. The roof of the whale's mouth contains a brushlike curtain through which the krill are sifted as the whale takes in vast gulps of water. The trapped shrimps are then swept backward into the narrow esophagus by the regular pulsating movement of the tongue.

The whale's whole gut is adapted and designed to receive a constant input of krill. The stomach is highly complex and very different to that of other carnivores. The first part of it is a croplike storage organ which regulates the supply of krill to the second chamber. This is a muscular unit that breaks up the tough, indigestible outer shells of the krill. In the final chamber, bile and pancreatic juices mix with the krill, reducing it to a watery chyme. Probably because the food is encased in so much inedible material, the baleen whale's intestines are exceptionally long — the longest found in any mammal. Intestines may be up to sixteen times the length of the body — thus the intestines of the blue whale may be as much as 1,600 feet long.

One of the least understood and most curious of all stomachs is that of the sperm whale. This whale — the original Moby Dick of Herman Melville's book — preys on giant squid, swallowing these huge mollusks alive. Inside the whale's stomach the squid is squeezed to death while enzymes set to work. In its death throes, the squid repeatedly attacks the stomach wall, which is protected from serious damage by a thick layer of a waxy secretion known as ambergris. Squid beaks become embedded in this layer, which is periodically shed and regurgitated. Ambergris has long had a fascination for man, for it possesses a strong earthy or woody odor which makes it a valuable additive in the manufacture of perfume.

The Herbivores

Man owes much to the existence of herbivorous mammals. Although these have lived since the earliest times of mammalian evolution, the spread of prairies in the Miocene epoch led to the explosive development of the even-toed ungulates, or artiodactyls, such as cattle and bison. There can be little doubt that the huge herds of artiodactyls

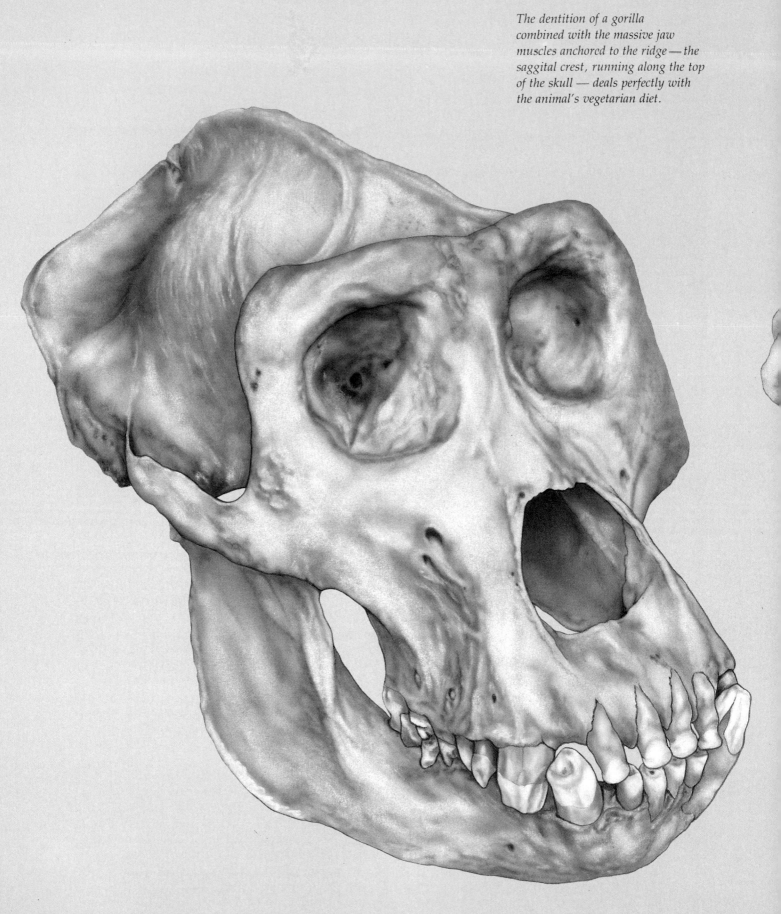

The dentition of a gorilla combined with the massive jaw muscles anchored to the ridge — the saggital crest, running along the top of the skull — deals perfectly with the animal's vegetarian diet.

That such a large creature as the southern whale can live on nothing but tiny krill — albeit taken in massive quantities — exemplifies another of nature's many adaptations to exploit available food.

A rodent such as this marmot (below right) has extremely hard incisor teeth for gnawing through tough vegetable matter — bark, leaves or stems. The teeth are self-sharpening.

which evolved during the Miocene were at least part of the evolutionary pressure which resulted in man's ancestors becoming gregarious and social, for only by hunting cooperatively could they kill these large and plentiful beasts.

The food of herbivores is made up largely of cellulose, but mammals are unable to make an enzyme capable of breaking down cellulose into sugarlike constituents. And the amazing aspect of the digestive system of herbivores is that mammals have entered into a symbiotic arrangement with a myriad bacteria and protozoa of various types which are able to manufacture cellulose-splitting enzymes. Bacteria are stored mainly in two parts of the gut — in a specially modified stomach and in a lateral pouch off the large intestine called the cecum. In cloven-hoofed ungulates, the bacteria are stored mostly in the stomach, while in that other major group of herbivores, the rodents, the bacteria occupy the cecum. The odd-toed ungulates, the horses and rhinoceroses, have bacteria in both the stomach and the cecum.

Plant material is tough. If you have ever tried to pluck grass by hand you will know that the silica cells in its blades can cut your skin to ribbons, so it is not surprising to find that herbivores have developed highly efficient plant-cropping mechanisms. Horses have large, strong, sharp incisors which neatly trim the grass. Cows and their relatives have no upper incisors; instead grass is cropped by the twisting and pulling action of the tongue coupled with the action of the lower incisors biting against a tough pad on the upper jaw. Being small animals, rodents have to gnaw through bark, stems and leaves, and to aid them in this, their single pair of both upper and lower incisors is coated on the outside surfaces with a thick layer of tough enamel. This wears away more slowly than the softer, inner parts of the tooth and retains a razor sharp edge.

It is essential that herbivores should have an efficient means of gathering food; since plant material is nutritionally far poorer than flesh or animal products, they must spend a great part of each day collecting food. The tiny field mouse eats its own weight of grass every day; the giant Panda,

*Because plant material is a poor
source of nutrients compared with
flesh, an herbivorous creature, such
as the manatee or sea cow, has to eat
vast amounts of its staple food, eel
grass, every day.*

a carnivore turned herbivore, spends ten to twelve hours of the twenty-four in feeding, and a manatee, or sea cow, fifteen feet long must consume eighty to a hundred pounds of eel grass daily. The elephant is probably at about the limit of size for a ground-feeding herbivore — if it were much larger it would require to feed for more than twenty-four hours each day to sustain itself.

Dealing with Cellulose

Evolution has provided three basic strategies for the digestion of cellulose, which are exemplified by the cow, the horse and the rodent.

The stomach of the cow is the most complicated known, for it comprises four distinct parts. If you watch a cow grazing, you will observe that it appears to bolt its food. It moves forward from one bunch of grass to the next, swallowing the food without stopping to chew it. For up to an hour it may do this, packing the largest section of its stomach—the rumen—with ripped up grass; it then retires and starts to chew the cud at its leisure.

The rumen is actually a fermentation chamber in which millions of anaerobic bacteria and protozoan ciliates start the process of cellulose digestion. Various organic acids, such as formic, acetic, propionic, butyric and lactic acids are produced in the rumen and are immediately absorbed into the blood. Muscular action of the rumen squirts small quantities of now softened grass back up the esophagus into the mouth, where it is chewed with the characteristic side-to-side jaw movement. It is at this point that the massive molar teeth are needed, for many strands of hard enamel run along the length of the teeth, providing a rasping action to the lateral jaw movement. The cud is chewed a few times on one side of the mouth and then a few times on the other.

By now the fibers are finely fragmented, and the food bolus is reswallowed. But this time it is channeled into the second chamber, or reticulum, directed to the correct place by the complicated muscle

20

action of the stomach wall. In the reticulum, water is pressed out of the bolus and absorbed for reuse in the rumen. Further water extraction occurs in the omasum, or third stomach, and the food then enters the abomasum, or true stomach. This is equipped with peptic glands, and here digestion actually occurs. The rest of the gut is little modifed, although it is typically quite long. In most cattle, the feces are very wet, but in species from arid regions, the rectum removes all the liquid available.

In horses a different strategy is employed. Horses are not ruminants, that is, they cannot chew the cud, and, once swallowed, food stays swallowed. This means that grass must be completely chewed when it is plucked. Horses have teeth just as good as those of cows for the job of chewing the cud but historically they have lacked the opportunity to retire to a safe place to do so. It is because they had to stand out on the open plain for far longer than the even-toed ungulates that they evolved into such fine runners, for they had to be always on the alert for wily predators. Indirectly, then, we owe the noble sport of horse racing to the peculiarities of the digestive system of the horse.

The horse's stomach is spacious and contains a huge quantity of bacteria as well as a number of protozoa; absorption of organic acids takes place here. In addition, the cecum is well developed, and much digestion also occurs there.

The rodents have developed a third strategy. In sharp contrast to the other two herbivore groups, the rodent has few bacteria in its essentially simple stomach, which serves as a holding organ where acids start to soften tough materials. The food passes fairly quickly into the small intestine and on into the large intestine with little digestion taking place on the way.

The cecum of rodents is huge, for it is here that the armies of bacteria are mustered. They attack the food in much the same way as those in the rumen of the cow, but now there is a problem, for the food is already past the part of the gut where most of the absorption of nutrients can occur. This is solved by the process of refection: as the food leaves the cecum to enter the final section of the large intestine, it does so in the form of soft pellets called cecotrophie. When these are expelled, the rodent takes them from the anus directly into the mouth.

The bison, so intricately a part of the history of the U.S.A., is a ruminant, a cud-chewing herbivore. To deal with the tough cellulose of grass, it has a complicated four-stomach gastric tract.

Now, on their second journey into the stomach, they pass into the cardiac limb where some further digestion occurs. Absorption of nutrients occurs as the cecotrophie pass along the small intestine. The now almost exhausted cecotrophie do not enter very far into the cecum, but some final absorption occurs there. Finally the waste matter is expelled in the form of small, dry pellets.

These three ways of solving the problem of breaking down cellulose demonstrate just how adaptable mammalian digestive systems are. The way they work may seem cumbersome, but that they are successful cannot be denied.

One fundamental question is, how do young herbivores acquire the bacteria in their gut? The answer is not known for certain, but young ungulates are often seen to eat soil shortly before weaning; perhaps this gives them an initial dose. Young koalas are known to be fed a special baby mash direct from the mother's anus just before weaning, and the same may occur in sloths. A small initial

The Australian honey possum (below) has a highly specialized mouth and tongue to gather up its staple food. But honey is so easily digested that the animal has only a simple stomach and gut.

Myth, legend and mystery surround the vampire bat (right). Attracted by warmth, it sneaks up on its prey and bites a hole in the victim's skin through which to suck blood. Blood, however, is such a dilute source of

nourishment that the bat must start urinating almost immediately it begins to feed if it is not to become too heavy to fly as a result of taking in a meal of blood sufficiently large to meet its needs.

dose of bacteria may be all that is required, since ideal conditions are provided in the rumen or cecum for the multiplication of microorganisms.

In calves, and presumably all ungulates, the rumen is very small at birth, even though later it will account for eighty percent of the total stomach volume. Mother's milk passes directly into the abomasum where the enzyme rennin is secreted and the milk coagulates. As the calf grows, the abomasum becomes overshadowed by the rumen and eventually accounts for only about seven percent of the total stomach volume.

Among the mammals there are a few highly specialized herbivores, but their adaptations mostly concern their food-gathering apparatus. Some of the better known are honey possums from Australia, and nectar- and pollen-feeding bats, which have long, feathery tongues for soaking up the sugary fluid and for dusting off the energy rich pollen. These high quality foods are readily assimilated and require only a simple stomach and gut, but their limited availability means that only the smallest of animals exploit them.

The bats are a huge and diverse group of mammals which contains one of the most bizarre of all mammalian life styles — that of the blood-drinking vampires. Blood is an excellent food source, being rich in proteins and other nutrients, but it is

difficult to handle because it is such a weak solution. It is all the more remarkable that the only mammals to exploit it should be bats which, like birds, cannot carry huge loads into the air. A vampire feeds for about two hours each night, consuming almost sixty percent of its body weight in fresh blood. Since this is far too much for it to fly with, the body starts to remove excess water the moment blood is ingested.

The vampire's stomach is a simple, thin walled tube, hardly worthy of the name. The blood passes directly from the esophagus into the duodenum; only occasionally will the stomach act as a storage vessel. Water absorption starts immediately and the efficient kidneys soon start to produce watery urine. Within three minutes of opening a wound on a prey animal the vampire starts to urinate. Two hours later it will have passed up to thirty-five milliliters of urine, weighing over one and a half ounces. As mentioned earlier, Masai tribesmen supplement their diet in the dry season with blood from their cattle, but this is no more than an emergency diet. If Count Dracula had existed, he would have had to drink over seven gallons of blood a day — or the blood of seven people.

Before leaving the subject of the digestive systems of man's mammalian relatives and considering the embryonic development of his gut, life in arid and hot places needs brief consideration. Carnivores consume a great deal of water with their food and are able to seek shade in the hottest part of the day. Their herbivorous prey, however, need to spend long hours feeding on relatively poor quality food, often far from shade in order to obtain sufficient water to meet their metabolic needs. Animals from arid zones rely for survival on having highly efficient large intestines which can extract every drop of water from feces — it is said that camel's dung can be burned the moment it is produced.

A further adaptation for desert life seen in camels is a mass of specialized water cells in the lower part of the rumen. These are goblet shaped structures which simply hold water, giving it up only when the fermentation chamber starts to run dry. Many a desert traveler has been saved from certain death by killing his camel and quenching his thirst on this reserve of water.

*Many a stranded traveler has been
kept alive by slaughtering a camel
and drinking the water in its rumen.
This water is stored there to
replenish juices involved in the
fermentation of food.*

*During development (right), simple
structures, based around repeated
folding of the embryo, grow to make
the organs that will eventually make
up the extensive and regionally
specialized digestive tract.*

Development of the Gut

Modern molecular biologists, having learned so much about the inner mechanisms of individual cells, whether those of bacteria, mice or men, are now beginning to grapple with second order problems such as the development of whole animals composed of billions of cells. This is an area of major mysteries. A single cell, the individual fertilized human ovum, contains one diploid set of chromosomes — forty-six in all. The physical structure of that cell, interacting with the genetic information residing in the DNA of the set of chromosomes, can make a complete human being.

In 1944, Erwin Schrödinger, an Austrian physicist, first clearly perceived that a chromosome contains molecularly coded messages that contain a specification for development. During the last forty years, a remarkably detailed picture of the language of those messages and the ways in which they are acted upon within cells has been built up.

But the underlying cellular mechanics of human development are only one way of seeking to understand this most fundamental of growth processes. Equally essential is an analysis of the physical and unfolding pattern of organ development itself, in the thirty-eight weeks of human intrauterine growth. Few organ systems other than the alimentary tract are so extensive, so regionally specialized

and give rise in embryology to so many other very distinct organs. The fact that the primordial gut in the first few weeks of our existence generates the beginnings of our middle ear, our tonsils, parathyroid, thymus and thyroid glands, lungs, liver and pancreas, should by itself be reason for seeking to understand this remarkable saga of human growth and development.

When a human embryo is about eight days old, the primordial zone that will eventually form almost all of the baby is unbelievably simple — it is merely a sandwich of cells with no "filling." The crucial cellular zone consists of a monolayer of ectoderm cells stuck to another layer of endoderm cells. By about fifteen days after fertilization, each of these cell layers abuts a fluid filled cavity. The ectoderm cells face the amniotic cavity, while the endoderm cells have extended around the inner lining of a space called the yolk sac.

It is this space and the cells that line it whose subsequent behavior explains so much about the embryonic derivation of the human alimentary tract. During the third week of embryonic life, ectodermal cells in the two-layered cell mass begin to move inward to form a "filling" for the cellular sandwich. These intucking cells form the mesoderm layer that will ultimately give rise to vital internal structures such as blood, bone, muscles and connective tissue. By the end of the third week, with mesodermal structures beginning to form, the previously flat embryo becomes more thoroughly three-dimensional by formation of head and tail folds. This folding of the cell layers also lays down the foundations for the human gut.

The portion of the yolk sac enclosed by the head fold will form all of the important gut structures from the mouth down to the level of the developing liver; this region is termed the foregut. The initially unenclosed central portion of the yolk sac will eventually construct the gut regions from the level of the liver as far down as a point which, in the adult, is in the colon. This is the midgut zone. The third and final zone, the hindgut, is that enclosed by the tail fold. This section forms the remaining, posterior, parts of the gut.

In the first few weeks of intrauterine life of the human embryo, a staggering series of changes occur in the foregut region. In order to com-

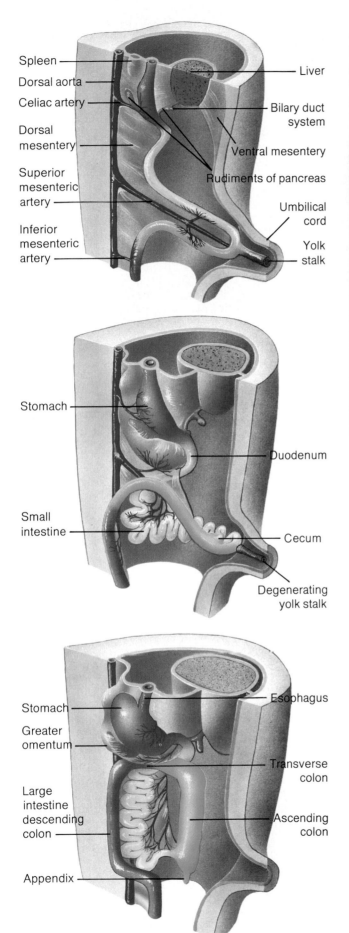

Spleen
Dorsal aorta
Celiac artery
Dorsal mesentery
Superior mesenteric artery
Inferior mesenteric artery

Liver
Bilary duct system
Ventral mesentery
Rudiments of pancreas
Umbilical cord
Yolk stalk

Stomach
Small intestine

Duodenum
Cecum
Degenerating yolk stalk

Stomach
Greater omentum
Large intestine descending colon
Appendix

Esophagus
Transverse colon
Ascending colon

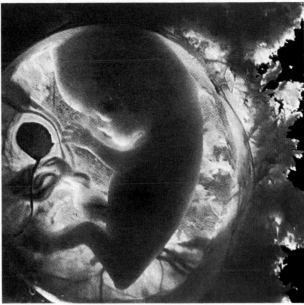

municate with the outside world, the foregut will need to link up with what will become the mouth. This starts off as a small dimple in the front end of the embryo which must join with the developing foregut. The dimple is tucked farther and farther in toward the foregut, which tunnels forward at the same time. Eventually the two link up, like the two halves of a tunnel being dug through a mountain, so that continuous communication is established. In the same way, the join between the two parts of the gut leaves no physical trace in the adult digestive system, but it is probably located just behind the gums.

During the fourth and fifth weeks of embryonic development, a set of four grooves, the pharyngeal pouches, appears along the side walls of the foregut just behind the invisible join. Simultaneously, corresponding ectodermal inpushings — the pharyngeal clefts — form on the outside of the embryo opposite each pouch. Pouches and clefts grow toward each other, but they normally do not merge. These strange repeating pits in the anterior part of the gut are the evolutionary remnants of the open gill spaces of our fishlike ancestors. In a fish they would become spaces for the ventilation of the gills in water; in man they are destined to become a collection of other, no less vital, organ systems.

The first pouch on each side will become the

The series of diagrams below illustrates in a left to right sequence the extraordinarily complex fates of the pharyngeal pouches and clefts of the human embryo. The first two images show the relatively simple *situation after five weeks of development, first in vertical section and then in the horizontal view, delimited by the dashed line in the left-hand illustration. The growth and migrations of cell clusters* *associated with the pouches are pictured in the right-hand pair of images, which show the developmental beginnings of thyroid, parathyroid and thymus glands, as well as that of the palatine tonsil.*

Five week embryo

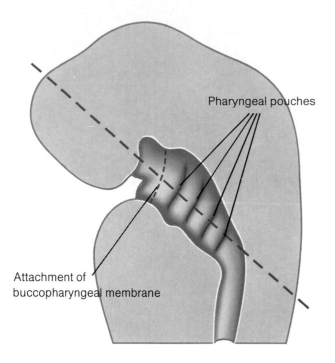

Pharyngeal pouches

Attachment of buccopharyngeal membrane

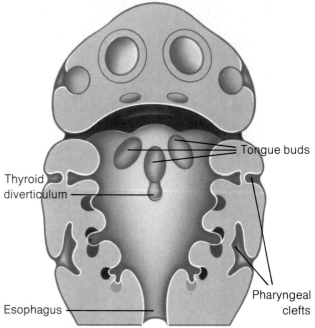

Tongue buds

Thyroid diverticulum

Esophagus

Pharyngeal clefts

middle ear, still connected to its gut origins by the Eustachian tube. The first clefts corresponding to these pouches become the outer ear canals, with the eardrums forming at the regions of contact between clefts and pouches. The remaining three outer clefts do not normally form adult structures —tissue between the first and second clefts grows backward obliterating these depressions. This overgrowth, or "burying," of clefts two, three and four forms the sides of the neck. In some children a rare congenital abnormality occurs in this region because of a failure of the precise control of cell movements. If overgrowth is not completely achieved, the child is born with a branchial fistula, which shows up externally as a hole in the skin on the side of the neck. The fistula leading inward from this hole usually ends in a lateral cervical cyst just above the angle of the jaw. Minor imperfections of this type show just how precisely coordinated and exact the constructive processes of embryology must be to achieve a perfect human being. The tiniest error in the earliest stages of body growth can have magnified final consequences.

Although clefts two through four will normally disappear, more positive fates await the remaining pharyngeal pouches. Four, rather than three, remain because the most posterior pouch is, in fact, subdivided into pouches four and five. Pouch two makes the primordium of the tonsil in the palate, and during the third, fourth and fifth months of development lymphatic cells move into it. Pouch three contributes cells to two vital glands; first, it makes the thymus, crucial for the processing of one important subpopulation of defensive white blood cells, the so-called T-lymphocytes. Second, it constructs part of the parathyroid gland, which controls aspects of calcium metabolism in the body. The closely associated pouches four and five also provide progenitor cells for glands. Pouch four joins with cells from pouch three to complete the primordium of the parathyroid gland, while the fifth pouch forms a small part of the thyroid gland termed the ultimobranchial (last-gill) body. The main part of the thyroid gland, manufacturer of the iodine-containing hormone thyroxin, is not formed from pouch material, but comes from ectodermal cells in the midline floor of the developing tongue as early as seventeen days into development.

Six weeks

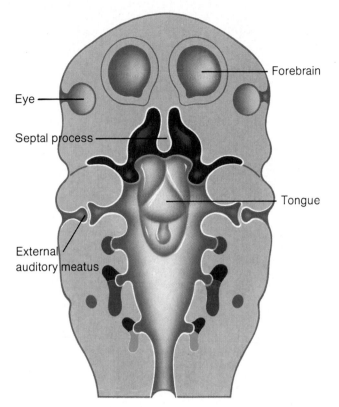

Eye

Septal process

Tongue

External
auditory meatus

Forebrain

Tongue

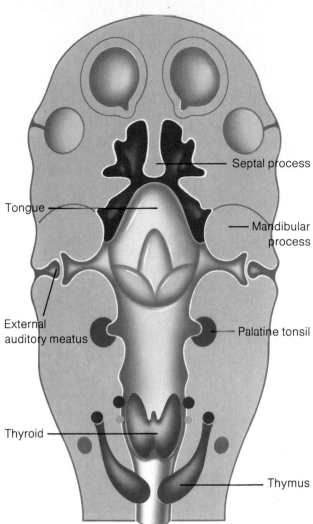

Tongue

External
auditory meatus

Thyroid

Septal process

Mandibular
process

Palatine tonsil

Thymus

These cells turn inward toward a deeper location and form the bulk of the thyroid gland.

Foregut tissue forms the lungs, esophagus, stomach, liver, gallbladder and pancreas. The esophagus and stomach are created from parts of the main foregut tube, while the lungs, the liver with its associated gallbladder, and the pancreas develop as outgrowths from the wall of the gut tube.

The midgut, which begins its existence with an extensive connection with the yolk sac, gradually reduces this linkage until it becomes a narrow vitelline duct. At the same time, it is elongating dramatically because it is destined to construct by far the longest section of the entire gut, namely, the hind part of the duodenum, the jejunum, ileum, cecum, appendix and about half of the colon.

Initially there is insufficient room in the ordinary peritoneal cavity to contain the expanding midgut, and it spreads out into the space within the umbilical cord during the sixth week of growth — only returning to a more orthodox position in the abdominal cavity at week ten, when more space is available. The great increase in the length of the gut in this zone is also the cause of the complex looping

that is characteristic of the small bowel.

The hindgut has the simplest series of changes during embryological growth, for it eventually forms the rear half of the colon, the rectum and the upper portion of the anal canal; the outer portion of the anal canal is an ectodermal inpushing. In addition, the hindgut contributes to the formation of the primordia of urogenital structures.

In developmental terms, we have now traveled from mouth to anus to understand how the tissue architecture of the gut is constructed. We have also seen how the digestive system in humans compares with those of other mammals.

It is now appropriate to take a closer look at the materials the human gut is designed to operate upon — human food — and see how the different components of the alimentary tract, such as the mouth and esophagus, the stomach and the small and large intestines, deal with the range of problems involved. Not only does the digestive system have to break down the food we eat into easily manageable units and extract all the nutrients necessary for healthy living, but it has also to dispose of the waste material that remains.

Chapter 2

Essential Fuels

The belief that you are what you eat is one that has been cherished throughout the ages. The mystical significance of food and drink is at the heart of many religions, especially Christianity, where the very deepest mysteries of the Church are expressed through the symbolism of bread and wine in the service of Holy Communion.

The food we eat is not only converted by the body into energy, and used for obvious physical activities encountered in everyday life, but is also needed for the constant process of growth and repair of body tissue. The three principal constituents of food are protein, fat and carbohydrate, which are all needed in varying degree in a healthy diet. Carbohydrates and fats are used to fuel all the body's processes and functions, while protein is mainly used as building material.

As well as these three basic components, the body must also have vitamins and minerals. Vitamins are essential for normal growth and development, and since they cannot be manufactured in the body, they must be supplied ready-made in the diet or, in certain cases, as supplements to the diet. Minerals assist in many body processes, such as normal nerve and muscle function, but are needed only in small quantities. A balanced diet nearly always provides sufficient quantities of minerals and vitamins and an excess of some vitamins can be harmful.

One of the most striking things about the human species is the wide range of foodstuffs on which it can survive and prosper. Many other mammals, such as gorillas, live on only a narrow range of plant foods. It is the capacity of humans to make use of all available food that led to the wide distribution of our ancestors throughout the globe, for, no matter where human beings are, they still seem to find sufficient nutrients to cover their needs.

Nutritionists tend to divide human societies into four main groups according to their pattern of nutrition. Our earliest ancestors belong to the first

The most important event of the year among rural populations, the harvest has been recorded many times on canvas, often allegorically as in this painting by Willem van Meiris (1662–1747), An Allegory of the Harvest.

29

group — the hunter-gatherers. In the slow development of early human species such as Australopithecus, the fossil humans, people ate large amounts of vegetable foods but gradually added to this vegetarian diet meat which they obtained by hunting.

Hunter-gatherers lived, as their name implies, from any food source they could find. However, their means of obtaining food, dictated a mobile, nomadic lifestyle. Once they had exhausted the supply of vegetable foods available in one area and denuded the area of game, they would move on and exercise their survival skills elsewhere. The way of life of the hunter-gatherer was — and still is — precarious, for their sources of food were unstable and they had nothing stored away. Today there are few people left on earth who practice this lifestyle; one notably successful group still survives though, the Bushmen in northwestern Botswana in southern Africa.

Discovery of Wheat

The birth of peasant agriculture, involving the cultivation of food plants, was the next evolutionary stage through which man's nutrition passed, and the next nutritional group emerged — those people who depend on the range of nutrients available from this form of food production. The group still survives throughout much of the world. The arrival of agriculture also made an enormous difference to the way in which human society was organized. The wandering life of the hunter-gatherer became settled, and villages and towns were built. This social evolution probably took place only because of the development, from ordinary meadow grass, of a strain of wheat that carried significant amounts of grain in its seedhead. The cultivation of these new strains of wheat probably started in the Middle East between 9000 and 8000 B.C.

The arrival of this new source of food brought with it a number of attendant nutritional problems. While hunter-gatherers seem to have a varied diet, there is a risk that a peasant agricultural community will depend upon one staple food, usually some sort of root vegetable or a cereal, and this can lead to nutritional deficiencies. Pellagra, a disease which results from a lack of the vitamin nicotinic acid, is perhaps the best example of a problem associated with a single staple food. It occurs almost exclusively in communities whose diet depends upon maize. Although there is plenty of nicotinic acid to be found in maize, the problem is that it is chemically bound up in such a way that the digestive system is unable to extract and absorb it. A further twist to this problem is the fact that the body itself is capable of making the missing vitamin from the amino acid tryptophan, but maize protein does not contain a sufficient quantity of this particular amino acid.

Another nutritional problem arises because a peasant agricultural community is at risk if the staple crop fails; the best-known historical example of this was the failure, in 1845 and 1846, of the Irish potato harvest, due to potato blight. In most circumstances, however, there is no shortfall of food as the seasons change, and a more settled lifestyle means that food can be stored from season to season.

The peasant agricultural community blends imperceptibly with the third main nutritional group — the affluent societies, who depend upon an efficient and industrialized agricultural system. It becomes clear that this third group has emerged because agriculture has become so organized and efficient that a relative degree of affluence and overproduction has been achieved. When this state of affluence is attained, the major nutritional problems become related to overeating, rather than undernutrition, and to difficulties such as excessive

When famine swept through Ireland in 1846, America raised one million dollars and sent relief ships. The aid program was mismanaged, however, and at least half a million people died of starvation.

But an excess of food can bring about its own problems. In some Western countries as many as one in five men and one in three women suffer from obesity and its related problems, (bottom).

consumption of alcohol, a drug which can readily be obtained from any grain or fruit crop.

Overpopulation, wars and droughts have led to a breakdown of established peasant agriculture in many parts of the world and this has led to the emergence of the fourth main nutritional group in the world today — the urban slum dwellers. These people have no immediate control over food production because they have no land on which to grow it and insufficient money to buy foods to meet their dietary needs. Although it is wrong to suppose that all the problems of this group are nutritional in origin, many of their problems appear to be the result of an inadequate diet.

The Science of Nutrition

Although the food we eat and the way we digest and use it has been such a major preoccupation throughout the ages, the science of nutrition was comparatively slow to develop. The great French scientist Antoine Lavoisier started to study energy utilization — an important aspect of nutrition — during the eighteenth century, and by the end of the 1780s he had completed a number of experiments on the amount of energy that animals burn up per day. However, it was not until the 1860s that further serious studies began into the way in which animals in general, and human beings in particular, change the food they eat into energy.

Under the patronage of King Maximillian II of Bavaria, the scientists Max von Pettenkofer and Carl von Voit laid the foundation of the modern understanding of nutrition. Nearly all energy production in the natural world depends upon the same basic chemical reaction: carbon- and hydrogen-containing compounds combining with oxygen breathed in through the lungs to produce water and carbon dioxide, which are excreted as waste products. Pettenkofer and Voit built an airtight chamber large enough for a man to live in for several days at a time. By measuring the amount of oxygen the man used, and the amount of carbon dioxide he breathed out, they were able to calculate how much energy was being used.

The American Wilbur Atwater, a student of Voit's in Munich, returned to the United States in 1892 and teamed up with the engineer Edward Rosa. Atwater and Rosa expanded on the Munich

Crowded into one room in a tenement building in New York in the early 1900s, with no land on which to grow food, urban slum dwellers were prone to illness brought about by malnutrition.

Today, however, food is more varied in America and health problems arise from too much food, or simply eating the "wrong" foods, (bottom). Fast food, in particular, has little to offer nutritionally.

work and built a chamber that enabled them to measure the amounts of oxygen consumed and carbon dioxide produced. They could also detect, to a high degree of accuracy, the heat energy produced by a man within their chamber.

The measurements that Atwater and Rosa made during the 1890s and the early years of the twentieth century form the basis of all subsequent studies of the energy values of the human diet. Once this pioneering work on the production and utilization of food energy was started, the whole subject area was thoroughly explored by the early 1920s. The important area of investigation then moved to an assessment of the difference in quality of the various nutrients contained within the human diet and what constituted a healthy diet.

In the years between the two World Wars, it was widely believed that a good, healthy balanced diet

Meat is a rich source of protein; however, the hanging carcasses in The Butcher's Stall *by Annibale Carracci (1560–1609) are a far cry from today's prepackaged presentation.*

High in carbohydrate and a source of roughage, wheat was once grown by peasant farmers as a means of survival; it is now big business, with many thousands of tons being exported each year (top right).

Pressing olives appears to be an extremely messy business (bottom right). *The oil extracted provides the body with energy and, in minute quantities, is used for growth and repair of tissue cells.*

relied on large amounts of energy and protein rich foods such as dairy produce, eggs and meat. Since World War II there has been a gradual realization that these foods may not be so healthy after all. Rather they may be exacting a major health penalty on the communities that rely on them by producing an increase in the incidence of diseases of the arteries resulting in problems such as heart attacks and strokes.

Most nutritionists believe that people in the West should be heading toward a greater reliance on carbohydrate foods. Although there is still no absolute agreement about the exact proportions of the three main classes of food — carbohydrates, fats and proteins — that the ideal diet should contain, it is widely accepted that a diet which supplies around fifty percent of its food energy in the form of carbohydrates is healthy. If this is so, then there are many people who should readjust their nutritional thinking, since a carbohydrate intake as low as thirty percent is not uncommon, with the twenty percent energy difference being made up by animal fats.

Dietary carbohydrates are exclusively vegetable in origin. Plants store their energy as carbohydrates and are harvested in the form of cereals, roots and fruits. Animals store little carbohydrate in their bodies. Most of the surplus energy they can extract from their diets is stored as fat. So a diet based on animal produce — lean meat, poultry or fish — is rich in the structural protein of animals and the animals's stored body fat. In contrast, it is unusual for plants to produce high yields of protein and fats, although there are exceptions. Some beans, especially the soya bean, contain large amounts of protein, and plants such as the olive, the sunflower and the coconut are exploited for their oil content. These exceptions aside, it is from plants that we get the first of the three main food classes, namely the carbohydrates.

Carbohydrates and Sugar

In current nutritional thinking, carbohydrates are the ideal energy source. Chemically, carbohydrates are combinations of large numbers of individual sugar molecules, and sugar is the body's preferred energy fuel. To a lay person, "sugar" is the substance produced by the sugar cane or sugar

This display of food (below) may look delicious, but it is a nutritionist's nightmare. It contains calorie laden food with a high fat content which has little part in a healthy diet.

This display of different types of bread, though less colorful, provides carbohydrates and also fiber (below right). Wholewheat bread is particularly high in fiber which aids the digestive system.

Methionine is one of the essential amino acids — an amino acid that the body is incapable of synthesizing. It is used as a dietary supplement and helps to hasten the removal of deposits of fat in the liver.

beet. To a scientist, sugars are a group of chemical compounds based on combinations of carbon, oxygen and hydrogen atoms.

Each of the simplest food sugars is made up of six carbon atoms, with six oxygen and twelve hydrogen atoms (six times H_2O), and glucose — the most important sugar in the human body — comprises six carbon atoms joined to six water molecules; hence "carbohydrate." Other simple sugars are formed when these atoms are joined in different ways. Compounds, such as fructose and galactose, have the same number of carbon, oxygen and hydrogen atoms as glucose, but the atoms are joined in a different configuration.

Glucose, fructose and galactose are known as monosaccharides, since each has the same basic backbone of six carbon atoms. It is when two of these six-carbon-atom sugars join that a disaccharide such as cane sugar is formed. The two monosaccharides that link to form cane sugar are glucose and fructose. It is monosaccharides, and glucose in particular, that the body needs for energy. The process of digestion splits the disaccharide sucrose into its two component monosaccharides, and little effort is required on the part of the digestive system for these monosaccharides to be absorbed and utilized.

Plants tend to store their energy in the form of the starch amylose, which consists of long chains of hundreds of glucose units. Starches are, therefore, called polysaccharides because they consist of many monosaccharide units. For the body to use amylose, the human digestive system has to break it down into individual glucose units. These will eventually be burned up to produce energy for the body's cells.

Although the body is unable to store large amounts of carbohydrate, there are limited stores available in both the liver and the muscles, in the form of the compound glycogen. Glucose units are joined together to make a glycogen molecule; there may be from two or three thousand up to sixty thousand. To be usable, any carbohydrate stored in the liver or muscles has to be soluble in water. The branched chemical structure of the glycogen molecule permits this, unlike that of amylose which is not water soluble.

Plant foods do not consist exclusively of digestible starch. They also contain varying amounts of other compounds, which the digestive system cannot cope with and which pass through the bowel relatively unchanged. The most important of these compounds is cellulose, which makes up the cell walls of plants. Cellulose and related indigestible compounds are often referred to as vegetable fiber or "roughage." Only since the mid 1960s has the importance of plant fiber to human health been appreciated. It was once regarded as

little more than an inconvenience, now it is realized that fiber is essential to the health of both the digestive system and of the whole body.

Proteins

The proteins are the second important class of major nutrients. Not only are they the body's essential building blocks, but they are also central to the way in which the body works, since all the thousands of chemical processes upon which human life depends are controlled by chemical regulators called enzymes. All enzymes are proteins.

In many important respects proteins are the keys to life. A one-hundred-and-fifty-pound man contains only about twenty-four pounds of protein. It is not a large proportion of the whole, but without it he would be unable to survive.

Proteins are characterized by the presence in them of the element nitrogen, in addition to the carbon, oxygen and hydrogen that make up the carbohydrates. A single protein molecule may contain thousands of atoms. Although proteins may have large and complex structures, they are all built up in a similar fashion: every protein consists of a string of quite small subunits, the amino acids.

In the entire animal kingdom there exist only twenty important amino acids. Each one of them has the ability to link up with any one of the other

nineteen: they are like the beads of a necklace which can easily be pulled apart and clicked back together in a new pattern. Each amino acid has a nitrogen-containing alkaline portion and a carbon and oxygen acid portion; it is these acid and alkaline pairs that join together to make up the string.

Twenty different types of beads might seem insufficient to make up the huge variety of different proteins. But the different amino acids are like the different letters of the alphabet. The text in this book is made up from words combining only twenty-six different letters. Few of the words contain more than fifteen letters, while a protein may contain thousands of amino acids — so the range of possibilities is enormous.

Each of the many hundreds of different proteins in the human body has its own role to play, either in the structure of the myriad different cells, in the huge number of chemical reactions occurring at any moment, or in the complex management of our hereditary mechanisms. The genes — units of DNA — that make up the chromosomes which produce the human blueprint, not only pass on the systems and characteristics common to all humans, but also the differences such as eye and hair colour that make us individuals.

Although plants do not contain as much protein as animals, the animal kingdom nevertheless depends upon plants to supply all its protein

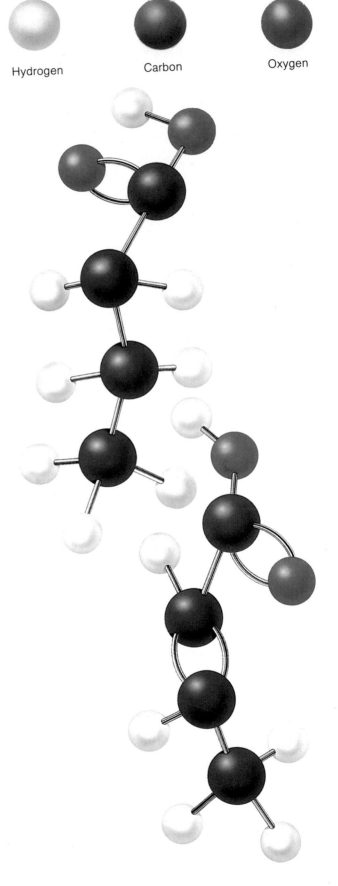

Examples of the three categories of macronutrients are glucose, a carbohydrate (below); saturated and unsaturated fats (right), which differ in the way the atoms are arranged, and a protein structure (far right).

Hydrogen   Carbon   Oxygen

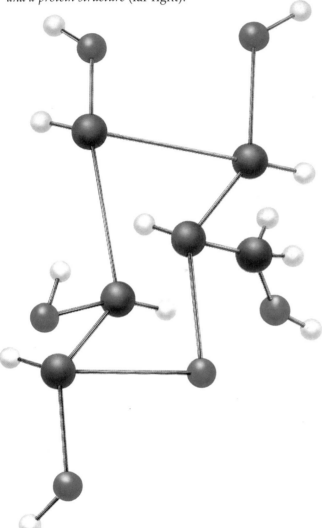

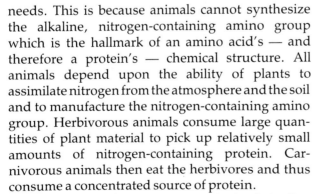

needs. This is because animals cannot synthesize the alkaline, nitrogen-containing amino group which is the hallmark of an amino acid's — and therefore a protein's — chemical structure. All animals depend upon the ability of plants to assimilate nitrogen from the atmosphere and the soil and to manufacture the nitrogen-containing amino group. Herbivorous animals consume large quantities of plant material to pick up relatively small amounts of nitrogen-containing protein. Carnivorous animals then eat the herbivores and thus consume a concentrated source of protein.

Animals must have some protein in their diet, but this dietary protein need not contain all twenty

Hydrogen Amino acid   Nitrogen   Carbon   Oxygen
          side chain

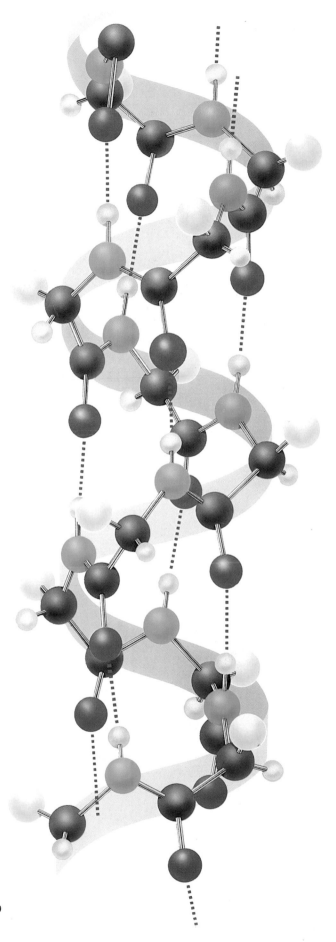

of the important amino acids. The human body can switch between many of the amino acids in order to produce the different combinations it needs. There are, however, a group of eight so-called essential amino acids which humans must have in their diet if they are to be able to synthesize the other twelve. These are leucine, isoleucine, lysine, methionine, phenylalanine, threonine, tryptophen and valine.

If the body takes in more protein than it needs for the growth and repair of its tissues, it can use some of the excess as a simple energy-producing fuel. The remainder is excreted, since the body has no way of storing protein. However, proteins are like many nutrients: you need an adequate supply to maintain health, but an excess is not necessarily good for you. It is quite difficult to take in large amounts of protein without relying on meat and dairy produce as major sources of supply, and if people eat large quantities of meat, they inevitably take in large amounts of animal fat.

The Fats — Energy Rich

Fats are the third main group of macronutrients — the nutrients that the body takes in in large amounts to provide it with energy and the raw materials for building.

Just as there is confusion between scientists and others about the word "sugar," so there is considerable confusion over the term fat. To most cooks, fat means a solid but easily meltable substance such as butter or margarine; oils, on the other hand, are seen as something slightly different. Chemically, butter, animal fat and vegetable oils all have a great deal in common. Certainly both the scientist and the cook would agree that the thing which particularly characterizes both fat and oil is the fact that they will not mix easily with water.

To remove some of the confusion over names, nutritional scientists tend to talk about "lipids," including all substances which are fatty or oily.

The body relies on lipids for two main purposes. First and most important is their structural role. Every cell in the body depends upon the lipid content of its cell membrane to preserve its functional integrity. Nutritional extremists who aim to banish all forms of fat from their diets would do well to remember this. Second, fat provides the most concentrated possible energy source within the diet. It

is, therefore, a very important nutrient which can turn into a major two-edged sword in affluent societies, where an excess rather than a lack of energy-producing food in the diet is the usual problem. When the body burns up one gram of fat, it produces a little over twice the energy yielded by a gram of protein or carbohydrate.

Just as starch is made up of glucose subunits, and protein is made up of amino acids, so there are basic subunits in dietary fat called fatty acids. Fatty acids are basically long strings of carbon atoms with hydrogen attached to them. Chemically, fats supply the same sort of energy source as gasoline, which is also made up of simple carbon strings with hydrogen atoms attached. However, the fatty acids differ from the hydrocarbons in gasoline in that they are usually longer strings, and they have a sort of chemical handle at one end in the form of an oxygen-containing acid portion. This is the same as the acid section of an amino acid. In the case of the fatty acids, the acid section is usually hooked up to a compound called glycerol. A single glycerol molecule has the capacity to hook up with three

fatty acids to make up a triglyceride, which is the form in which fats usually occur in foods.

Unlike the carbohydrates and proteins, the lipids have a wider range of naturally occurring structures. They may bind together with phosphorus to form phospholipids, which are important in the structure of cell membranes. There are also highly important lipids, called sterols, where the chains of carbon atoms form rings. Perhaps the best known of these is the compound cholesterol, which has been closely investigated because of its association with heart and artery disease.

It seems that high levels of cholesterol in the blood are associated with a greater incidence of heart and artery disease. This finding was suspected for many years but has unequivocally been shown to be so by the huge study of the population of Framingham, Massachussetts, masterminded by Professor William B. Kannell of Boston University Medical Center.

Most of the cholesterol in the bloodstream is made in the liver; only about twenty percent is derived from the cholesterol taken in as food.

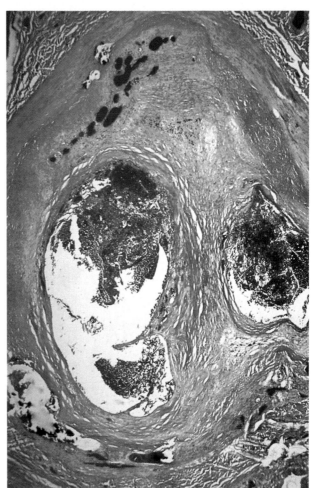

In atherosclerosis (right), fatty streaks appear on the inner wall of the arteries. These streaks can become a hard mass of fatty tissue, which erodes the wall and restricts the flow of the blood.

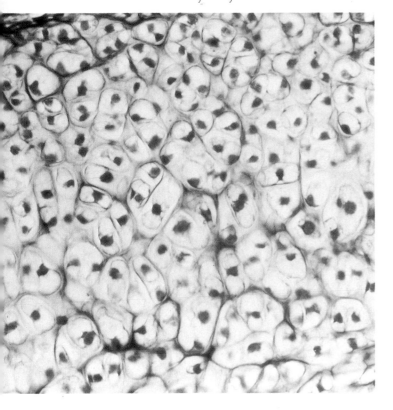

Nutritionists now believe that the best strategy for lowering the general level of blood cholesterol in the population is to try to decrease the amount of fat contained in the diet.

Saturated Versus Unsaturated Fat

The terms saturated and unsaturated fat are freely used by media personnel, and the lay public, but many people fail to grasp what they actually mean. The terms refer to the chemical structure of the fatty acid chains that make up the triglyceride in dietary fat. If a carbon atom in a carbon chain is linked to two hydrogen atoms, then it is using up the whole of its capacity for chemical linkage. This is called saturated bonding. However, it is also possible for the same atom to link with only one hydrogen atom; this is unsaturated bonding. The fatty acids which plants make often contain a number of these unsaturated bonds; the fatty acids in animal fat, on the other hand, are usually constructed with fully saturated bonds.

It seems that if the diet contains a high level of fat of animal origin, that is, a high level of "saturated fat," there is a greater tendency for the liver to manufacture cholesterol, which leads to a higher level of cholesterol in the blood. This in turn leads to an increased risk of heart and artery disease. If the total level of fat consumed is reduced, and the proportion of unsaturated to saturated fat is increased, then the level of cholesterol in the blood should be lowered.

The epidemics of heart and arterial disease that have swept through the developed nations since the end of World War I are due, at least in part, to inadequacies in nutrition. Huge amounts of time and money have been expended in trying to combat the problem. As a result there will almost certainly be a dietary shift away from fats and toward carbohydrates as major providers of energy — or macronutrients — in the diet.

It often takes a disease of major economic importance to bring about a shift in dietary thinking, as the story of the micronutrients illustrates. Micronutrients are the vitamins and minerals, which the body needs in only small amounts but which are essential to health. The big expansion in human

James Lind

Saviour of Maritime Nations

James Lind, generally acknowledged as the founder of nautical medicine, was born in Edinburgh, Scotland in 1716. His family's medical connections led to his apprenticeship in the field at the age of fifteen. Eight years later, he entered the navy as a surgeon's mate and for the next ten years served in the West Indies, the Mediterranean, the coast of Guinea and the English Channel.

A naval career for a common sailor was hard and often short. Apart from the usual casualties of war, illnesses of all kinds were endemic on board ship. The biggest scourge was scurvy, which was capable of killing off more than half a ship's crew on a long voyage. From around 1500, when oceanic sailing became practicable, foodstuffs had to last for many months at sea and so were salted, smoked or dried. The lack of vitamin C in preserved food, together with the lack of fresh foodstuffs, led inevitably to outbreaks of scurvy at any time between two to eight weeks into a voyage, since the vitamin can be stored for only a very short time in the body.

It had been known for two centuries that scurvy could be cured by administering orange or lemon juice, but it was Lind's experiment and subsequent publishing of his findings which proved conclusively that ascorbic

acid was a specific against the disease. In his book *A Treatise on the Scurvy* (first published in 1753), he describes how he selected twelve sailors suffering from the illness. For six days, he administered different treatments to the six groups, ranging from sea water to citrus fruits.

Those being treated with citrus fruit showed the greatest signs of recovery, one even becoming sufficiently well to nurse his fellow sufferers. Furthermore, Lind's book recommended ways of storing citric acid and other antiscorbutics such as onions and sauerkraut (used with success by the Dutch navy and later by Captain Cook). He also suggested adding fruit wines to ships' stores, growing small green vegetables on deck and using provision ships to take fresh food to the fleet.

Lind's treatise went through three editions in English and one in French. The interest aroused by its publication was eventually instrumental in the enforced issue of lemon juice in all ships. By 1800, scurvy had been almost wiped out; and not a moment too soon. The British navy was now strong enough in manpower to protect England from invasion, sweep the French from the seas and thus contribute to the downfall of the Napoleonic regime.

Though chiefly known for his work on scurvy, Lind was concerned with all aspects of naval health. His *Essay on the Most Effectual Means of Preserving the Health of Seamen in the Royal Navy* dealt with the problems of ventilation, overcrowding and bad water supplies. A third book, entitled *Essay on Diseases Incidental to Europeans in Hot Climates*, gives acute deductions founded on close observation of tropical diseases such as malaria, sleeping sickness and tetanus.

In 1757, Lind was appointed physician in charge of the newly founded Haslar Naval Hospital at Portsmouth which he headed for twenty-five years. Most of the recommendations he had urged in a lifetime's involvement with nautical medicine laid foundations that were radically to improve the lot of the common sailor in the years following his death in 1794.

knowledge of the micronutrients depends upon the advances made in analytical chemistry in the second half of the nineteenth century. The whole concept of micronutrients did not really emerge until the turn of the century, but practically all the important substances had been identified by 1930. However, the roots of the vitamin story lie well back in history, when scurvy, an often fatal disease, emerged as a major problem, particularly for the maritime nations.

Scurvy — the Sailors' Curse

Scurvy is an incapacitating disease which causes prostration, bleeding disorders and eventual death. It became of economic importance to the European nations following the sacking of Constantinople in 1453. When the Turks took the city, it not only marked the final gasp of the Roman Empire but also cut off the overland trade routes to the East. This event catalyzed maritime exploration and set Europe on its seaborne expansion.

In 1497, the Portugese explorer Vasco de Gama successfully sailed around the Cape of Good Hope and established a trade link with the East. Two-thirds of his men died of scurvy on this voyage, and, for the next three centuries, scurvy was the constant companion of the long-distance mariner. Although the great French explorer Jacques Cartier was shown how to cure the disease by Indians in Newfoundland in 1535, no permanent method of prevention or cure resulted.

The man who was responsible for providing a reliable and specific preventive approach to scurvy, and a reliable cure for sufferers, was the British naval surgeon James Lind. What Lind did was to compare the effects of different types of treatment on a group of people suffering from the disease. Although it seems an obvious approach, this critical experiment, performed by Lind in 1747, was the first example of what we would now recognize as a controlled trial of treatment. This sort of comparison has since become the basis of the therapeutic revolution of the twentieth century.

VITAMIN	BEST SOURCES	ROLE	RECOMMENDED DAILY INTAKE
Vitamin A (Retinol)	Liver; milk; eggs; butter; dark green or yellow fruits and vegetables. The body converts the pigment carotene in yellow and green fruit and vegetables to vitamin A.	Needed by body membranes, including the retina of the eye, linings of lungs and digestive system. Also needed by bones and teeth.	about 1 mg
Thiamin (Vitamin B$_1$)	Pork; whole grains; enriched flour and cereals; nuts; beans and peas	Ensures proper burning of carbohydrates.	1.0-1.4 mg
Riboflavin (Vitamin B$_2$)	Milk; cheese; eggs; liver; poultry	Needed by all cells for energy release and repair.	1.2-1.7 mg
Nicotinic Acid	Whole grains; enriched flour and cereals; liver; poultry; lean meat	Needed by cells for proper use of fuel and oxygen.	13-19 mg
Pyridoxine (Vitamin B$_6$)	Liver; lean meat; whole grains; milk; eggs	Needed by red blood cells and nerves for proper functioning.	about 2 mg
Pantothenic Acid	Egg yolk; meat; nuts; whole grains	Needed by all cells for energy production.	4-7 mg
Biotin	Liver; kidney; egg yolk; nuts; most fresh vegetables	Needed by skin and circulatory system.	100-200 micrograms
Vitamin B$_{12}$	Eggs; meat; dairy produce	Needed for red blood cell production in the bone marrow. Also needed by nervous system.	3 micrograms
Folic Acid	Fresh vegetables; poultry; fish	Needed for red blood cell production	400 micrograms
Vitamin C Ascorbic Acid	All citrus; tomatoes; raw cabbage; potatoes; strawberries	Needed by bones and teeth and by tissues for repair.	60 mg
Vitamin D	Oily fish and fish liver oils; dairy produce; eggs	Required for maintenance of blood calcium levels and thus for bone growth. Some vitamin D can be made in the skin in the presence of sunlight.	5-10 micrograms
Vitamin E (Tocopherol)	Vegetable oils and many other foods	Needed for tissue handling of fatty substances and for making cell membranes.	8-10 mg
Vitamin K	Made by intestinal bacteria; found in leafy vegetables	Needed for normal blood-clotting.	70-140 micrograms

MINERAL	BEST SOURCES	ROLE	RECOMMENDED DAILY ALLOWANCE
Calcium	Dairy produce; green vegetables	Essential for blood-clotting and the structure of bones and teeth. Needed for working of nerves and all other electrically active body tissues.	about 800 mg in adults but more during growth
Phosphorus	Meat; dairy produce; beans and peas and cereals	Basic cell energy store; key element in cell reactions.	about 800 mg in adults but more during growth
Potassium	Avocados; bananas; apricots; potatoes and many other foods	Major mineral within body cells. Essential to fluid balance and for many cell reactions.	about 3 g
Magnesium	Beans and peas; nuts and cereals; leafy green vegetables	Needed by all cells. Important in electrical activity of nerves and muscles.	up to 500 mg
Iodine	All seafood; iodized salt	Needed by thyroid gland.	about 0.1 mg
Iron	Liver; meat; eggs; enriched cereals	Needed in manufacture of hemoglobin, the oxygen-carrying compound in blood.	10-15 mg
Fluorine	Water; fluoride toothpaste	Helps protect teeth from decay.	–
Copper	Liver; seafood; meat	Needed by cells to utilize oxygen.	about 1.5 mg
Zinc	Seafood; meat; wholewheat; beans and peas; nuts	Needed in the structure of cell enzymes.	15 mg
Chromium Selenium Molybdenum Manganese	Trace elements in many foods	Minor roles in body chemistry.	minute amounts

This Balinese painting Harvesting the Rice, *presents a pleasant picture, but in the late 1800s, polished rice, stripped of its nutritious husk, had a devastating effect on the Japanese navy.*

Lind was on active service on HMS *Salisbury* and had twelve sufferers from scurvy under his care. Dividing them into groups of two, he instituted treatment with six different regimens. He used cider for one group, a weak elixir of sulfuric acid for another and vinegar for a third. Two other groups were given seawater to drink or a curious concoction of nutmeg, garlic, balsam of Peru and myrrh. None of these five pairs improved. Only the fortunate sixth group, the two sailors who were given two oranges and one lemon every day, showed an improvement; in these two, the effect was dramatic, so much so that one of them was back on duty within a week.

New ideas take a long time to get accepted even now, and they took even longer in the eighteenth century. Lind's treatment might never have been taken up had not the great Captain Cook believed in it. On his circumnavigation of the globe, which lasted from 1772 until 1775, he was able to prevent scurvy breaking out among his crew by making sure that he took on fresh fruit and vegetables wherever possible. In 1795, during the Napoleonic wars, the Lords of the Admiralty finally accepted that fruit could prevent scurvy in the crews of their men-of-war, and it became the rule that British sailors were given citrus fruit daily. This had the immediate effect of doubling the fighting strength of the British navy, since it was commonplace in those days for half the crew to be laid up with scurvy on a long voyage. The British also earned their nickname of "limeys" in this way.

It is now known that scurvy results from a lack of vitamin C, whose correct chemical name is ascorbic acid. This substance is plentiful in raw fruit, especially citrus fruit, tomatoes and rose hips, but much of it is destroyed if the fruit is cooked.

Rice, Beriberi and Vitamin B

One hundred years after the introduction of fresh fruit into the diet of English sailors, the Japanese navy was being laid low by beriberi, a disease which causes tiredness, weight loss and heart failure. In 1879, nearly two thousand of the five thousand men in the Japanese navy were off active service with beriberi and fifty men died from it. As a result of the efforts of another naval surgeon, named Takaki, the disease had been eliminated within ten years. Takaki thought that there was too great a proportion of polished rice in the rations of the Japanese sailors. He persuaded the authorities to replace the rice with other foods and succeeded in preventing the disease, but he mistakenly attributed his success to a higher level of protein in the new diet.

The next step in the discovery and recognition of the B vitamins was made in 1890, by a military physician, this time an internist named Christian Eijkman, who was working at the Dutch military hospital in Java. He had plenty of beriberi victims under his care, and he had the novel idea of feeding their normal hospital diet to some chickens. The chickens soon developed weaknesses, but at this point the hospital cook discontinued Eijkman's supply of special military polished rice and he had to make do with ordinary whole rice for his chickens. Oddly enough, the birds got better . . . . Eijkman spotted the significance of this chance act of military parsimony and went on to extract a factor from rice-washings which could prevent chickens suffering from beriberi. For this discovery he won a Nobel prize.

The essential element found in the rice germ is now known to be thiamine, or vitamin B_1. The B-vitamin group consists of eight different compounds. These eight, along with vitamin C, make up the nine water-soluble vitamins. There are also four vitamins which cannot be dissolved in water but which will dissolve in fats and oils. These are the fat-soluble vitamins: A, D, E and K.

Vitamin A was not found by dietary experiment in response to an economically important scourge like scurvy. The discovery of the vitamin resulted from the work of Sir F. Gowland Hopkins, Professor of Biochemistry at Cambridge University, England, in the first decade of this century. Interested in the effects of milk, he fed young rats on a diet which seemed to contain an adequate range of macronutrients and sufficient energy, but no milk; the rats died. But when he added milk to the diet of another group of rats, they thrived. Hopkins had demonstrated the presence of an accessory food factor in milk, and thinking that this was one of the amine group of compounds, he coined the phrase "vital amine" or vitamin.

Hopkins was instrumental in establishing the importance of micronutrients in general and vitamins in particular. However, it was left to two pairs of American workers to isolate this vitamin in butter, eggs and cod-liver oil, in 1913. Later, one of these pairs, E.V. McCollum and N.S. Davis, went on to name the compound "fat-soluble vitamin A" to distinguish it from the "water-soluble vitamin B," which, following on from the work of Eijkman, they had found in polishings from cereals. (We now know that "vitamin B" actually consists of a group of eight different compounds.) The distinction which they made between the water- and fat-soluble vitamins is an important one, since the two sorts of vitamin are handled very differently by the absorption mechanisms in the gut. In developed countries the diseases of vitamin deficiency now occur only in people with abnormal diets, or, more commonly, in people who have diseases that interfere with their absorption of food.

The recognition of the other vitamins followed on

quickly from the characterization of vitamin A. In 1918, the English worker Mellanby showed that there was a fat-soluble substance in cod-liver oil which prevented rickets in puppies. Pupils of Hopkins were able to isolate vitamin C, which was finally synthesized artificially in 1932.

In 1923, the experiments of Evans and Bishop in California led to the recognition of vitamin E. These workers showed that the proper reproductive function of both male and female rats depended upon a fat-soluble substance, which they later identified and called tocopherol, or vitamin E. The discovery of vitamin K, one of the factors upon which blood-clotting depends, followed from Denmark in 1934, and the final discovery in the vitamin story was made when the last of the B vitamins, vitamin B_{12}, was simultaneously isolated in Britain and in the United States in 1948.

Vitamins, Minerals and Health

The vitamins are not the only micronutrients needed to maintain health. There are also a number of important minerals, such as calcium, iron, phosphorus and iodine, upon which the body depends and which may have a number of different roles to play. Sodium, potassium and magnesium are essential to the functioning of all the cells, as are calcium and phosphorus, which have a separate function, since they are integral parts of the structure of bone. There are also other minerals, which may not be needed in particularly large amounts but which control a single vital function within the body. Iron is essential for the formation of hemaglobin, the blood's oxygen carrier, for example.

Vitamin A, necessary for the formation and maintenance of the skin, is supplied by yellow and orange vegetables such as carrots (below) which contain carotenoid pigments. And sunshine (right)

provides vitamin D, which increases calcium absorption from the gastrointestinal tract and is essential for normal bone growth. A deficiency of vitamin D results in the distorted bone formation of rickets (bottom).

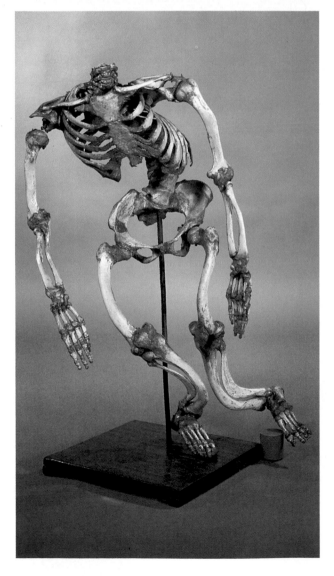

The important aspect of both the vitamins and the minerals is that they are needed in only small amounts in the diet. In recent years, there has been a steep rise in the use of extremely large amounts of vitamins in the mistaken belief that "if some is good for you, more must be better." This approach is potentially dangerous nutritionally, an excess of vitamin A can cause insomnia, weight loss and aching limbs, and there have been a number of deaths as a result of overdosage of both vitamin A and vitamin D. A normal mixed diet will supply both the macronutrients and the micronutrients that are essential for health and excess of any of these may lead to more harm than benefit.

Chapter 3

The Journey Commences

The section of the digestive system from the mouth to the start of the duodenum is the upper gut. It marks the first stage of the long journey that food takes from the moment of eating to the time all the nutrients have been extracted by the complex process of digestion. During this first part of the journey, food is transformed into an even, homogenized liquid. After being ground by the teeth and lubricated by saliva, it is then swallowed and churned in the acid broth of the stomach. Constant squeezing and pushing by the muscular walls of the gut keeps the production line of digestion on the move, and specialized cells in the lining of the upper gut manufacture the chemical brew that marinates the tough protein in the food, ready for breakdown and absorption which take place in the small bowel.

With all this complicated processing going on, it is not surprising that mild discomforts of indigestion occur. Also, the more serious conditions of peptic ulcer and gastric cancer are reminders of how sensitive this part of the gut is to disease.

The Mouth

More than just the entrance to the gut, the mouth serves in many ways in breathing and talking and in aspects of social and sexual behavior. It is the only part of the gut that is easily open to inspection and is important to doctors as an indicator of disease. The lips, with their sharp, clearly defined "bright red border" mark the start of the mucous membrane that lines the whole of the inside of the gut. Under this membrane is a circular muscle, called the orbicularis oris, which is tagged onto the muscles of the face and guards the access to the mouth. The orbicularis oris also permits all mouth movements, from a trace of a smile to a big grin, as well as opening the lips to eat. When drinking, the lips are coordinated by this muscle, allowing us to sip a liquid, delicately scooping it in small amounts from a glass or cup.

The lady placing a sandwich in her mouth in Edward Burra's Snack Bar *is probably unaware of the complexities of the digestive process. These are initiated by the act of eating.*

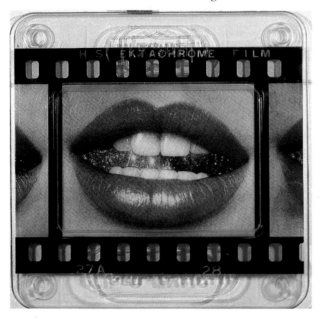

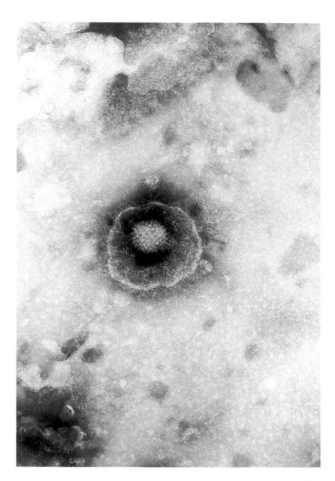

In the womb, the development of the lips and the front part of the mouth occurs through the fusion of a central and two side outgrowths from the developing head. Failures in this pincer movement in the growing fetus cause the common deformity known as "hare lip." When it occurs farther back in the mouth, a cleft palate develops. Causes are unknown, but some suspicion is drawn to environmental influences in the first three months of pregnancy, including exposure to X rays, cortisone drugs, vitamin deficiency and virus infections in the mother (especially rubella, or German measles). In about one case in ten, however, the condition is hereditary, and sometimes it accompanies other more serious malformations of the head.

Surgical repair is essential, not only to prevent the emotional scars that the facial deformity will bring but also to ensure the correct alignment and use of the teeth. Hare lips are usually repaired in the first year of life, with careful suturing to realign the lips, in an operation known as a V-Y repair from the shape of the incisions made. If the palate is also cleft, repair is essential to aid clear speech. However, repair of the palate is a more intricate operation and is not usually attempted until well into the second year. It is a considerable relief to parents, since open cleft palates cause many feeding difficulties and result in discomfort for the baby.

In adult life, one common problem with lips is cold sores, which often occur in strong sunlight or accompany some virus infection such as flu. The story normally starts in childhood, when a more severe eruption of blisters and sores in the mouth is caused by the herpes simplex virus. Although this settles after some days or weeks, the virus particles do not depart entirely but secrete themselves in the nerves of the face, reappearing in sunlight or when some other infection lowers the body's immune resistance to the herpes simplex virus. Small blisters then appear around the mouth, often at the junction of the skin and lips: these are what are termed "cold" sores.

Inside the lips, the cavity of the mouth is formed by two bony outgrowths of the skull. The upper jaw (maxilla) is a direct outgrowth of the lower part of the skull, while the lower jaw is joined by a hinge, just in front of the ears, to the bones of the temple. Powerful muscles attached to the lower jaw make the mouth and teeth into cutting pincers or meat grinders that can tear and grind food. More subtle changes in the aperture of the mouth are needed to produce the sounds of speech and open-mouthed smiles.

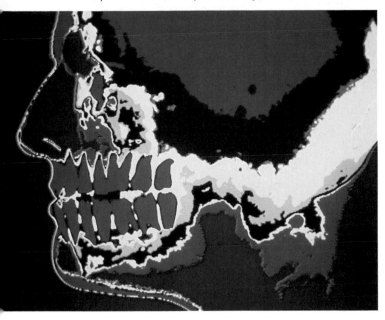

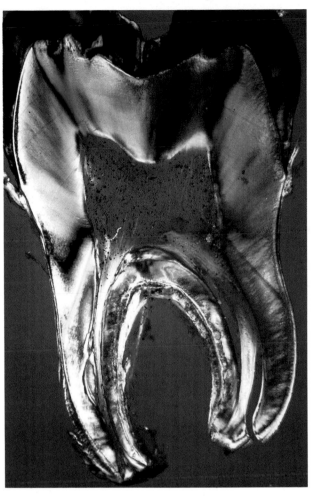

Human teeth are mounted firmly in special outgrowths arising from the upper and lower jaws known as the alveolar processes and are cushioned by the gums. The gums are firmly attached to the bone of the alveolars and are covered with a pink, velvety mucous membrane. This, like the membrane lining the rest of the mouth, contains small mucus-secreting glands which keep the mouth moist. The teeth are coated on the outside with a tough enamel layer and are made principally of a hard, bonelike material, dentine. Within the center of each tooth are blood vessels and nerves, known as the pulp. This is lined on the outside with a layer of cells, the odontoblasts, responsible for manufacturing the overlying dentine. The enamel coating of the teeth is made by further specialized cells, ameloblasts, which disappear once the enamel has been applied to the growing teeth and are no longer found in mature teeth.

Growth of the Teeth

The intricate structure of the teeth is not "anatomical overkill." Once past the age of twenty or thereabouts, we can expect no replacements from nature, so the teeth must be strong and durable.

We are all supplied with two sets of teeth, the milk, or deciduous, teeth and later our permanent dentition. Despite the fact that we may not see our permanent teeth until midchildhood, many of these start their development in the womb and by birth are largely formed, waiting their turn behind the first set of temporary teeth. As early as the third month of pregnancy, the buds of the permanent teeth have formed in the fetal jaw. Neither set of teeth acquires its roots and becomes firmly attached to the alveolar processes until just before it erupts. In most people, the milk teeth make their appearance during the first year of life. It is unusual for a baby to be born with teeth already erupted, but this does occur in about one in two thousand children, including (it is reputed) Julius Caesar, Hannibal and Napoleon. In some cultures, children born with teeth were viewed with superstitious distrust, and once they were even feared and killed as devils in some parts of Africa.

51

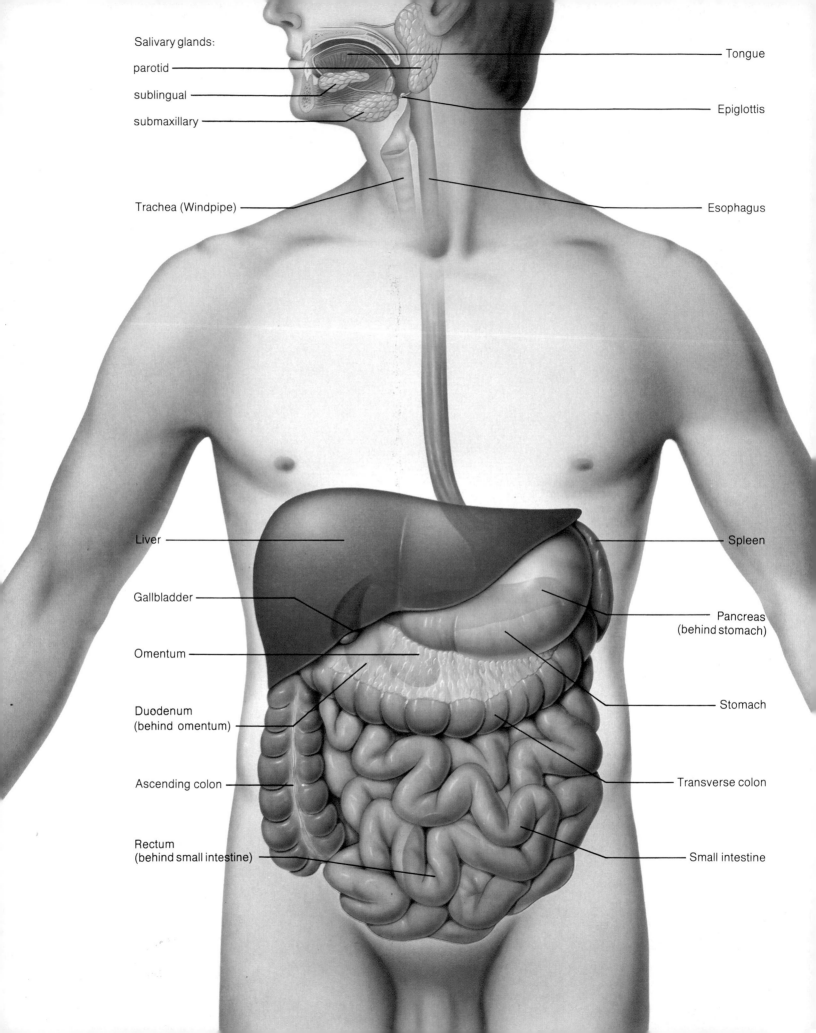

Salivary glands:

parotid

sublingual

submaxillary

Trachea (Windpipe)

Liver

Gallbladder

Omentum

Duodenum
(behind omentum)

Ascending colon

Rectum
(behind small intestine)

Tongue

Epiglottis

Esophagus

Spleen

Pancreas
(behind stomach)

Stomach

Transverse colon

Small intestine

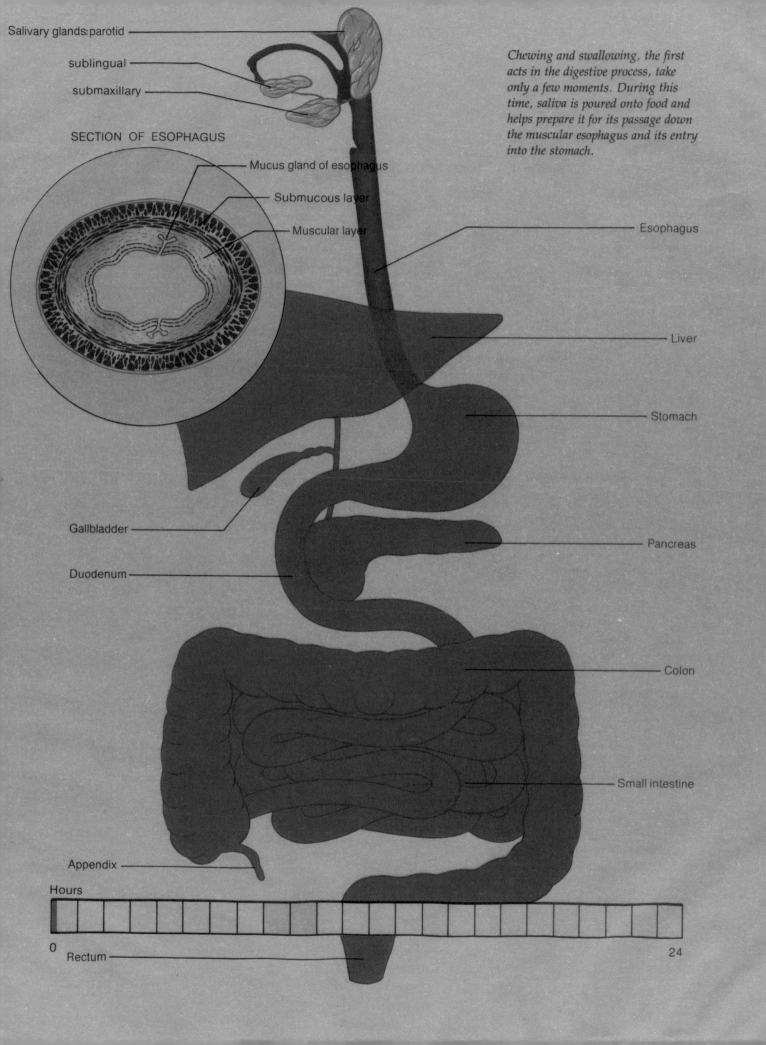

Salivary glands:parotid

sublingual

submaxillary

SECTION OF ESOPHAGUS

Mucus gland of esophagus

Submucous layer

Muscular layer

Chewing and swallowing, the first acts in the digestive process, take only a few moments. During this time, saliva is poured onto food and helps prepare it for its passage down the muscular esophagus and its entry into the stomach.

Esophagus

Liver

Stomach

Gallbladder

Pancreas

Duodenum

Colon

Small intestine

Appendix

Hours

0   24

Rectum

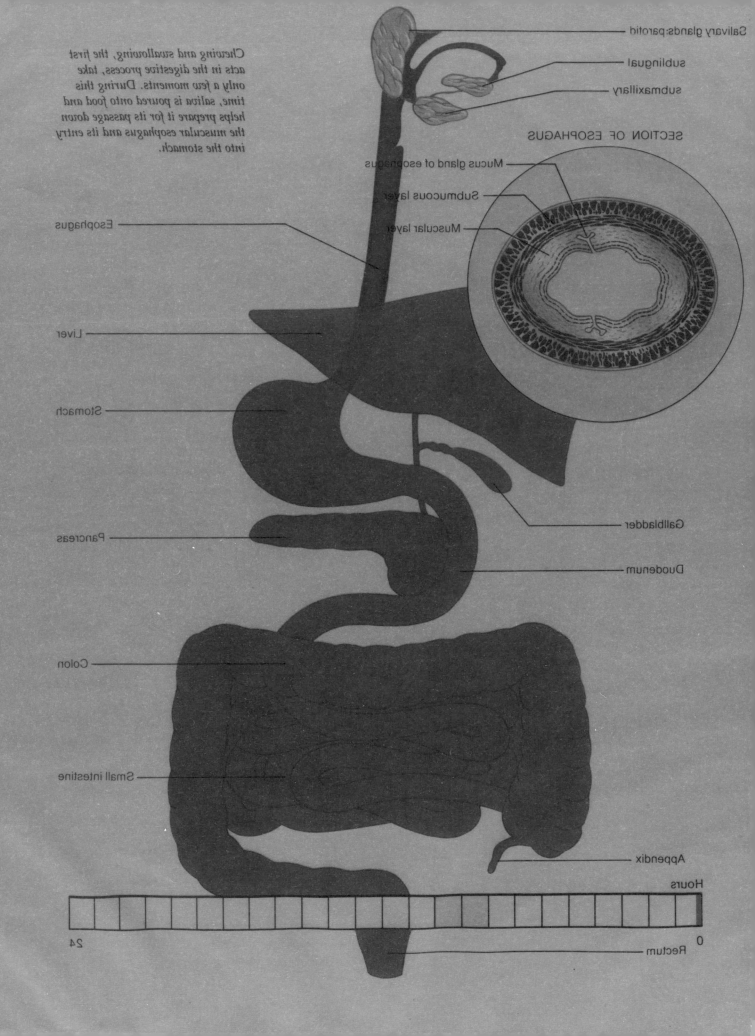

Liver
Stomach
Pancreas
Gallbladder
Duodenum
Colon
Small intestine
Appendix
Hours
Rectum

24
0

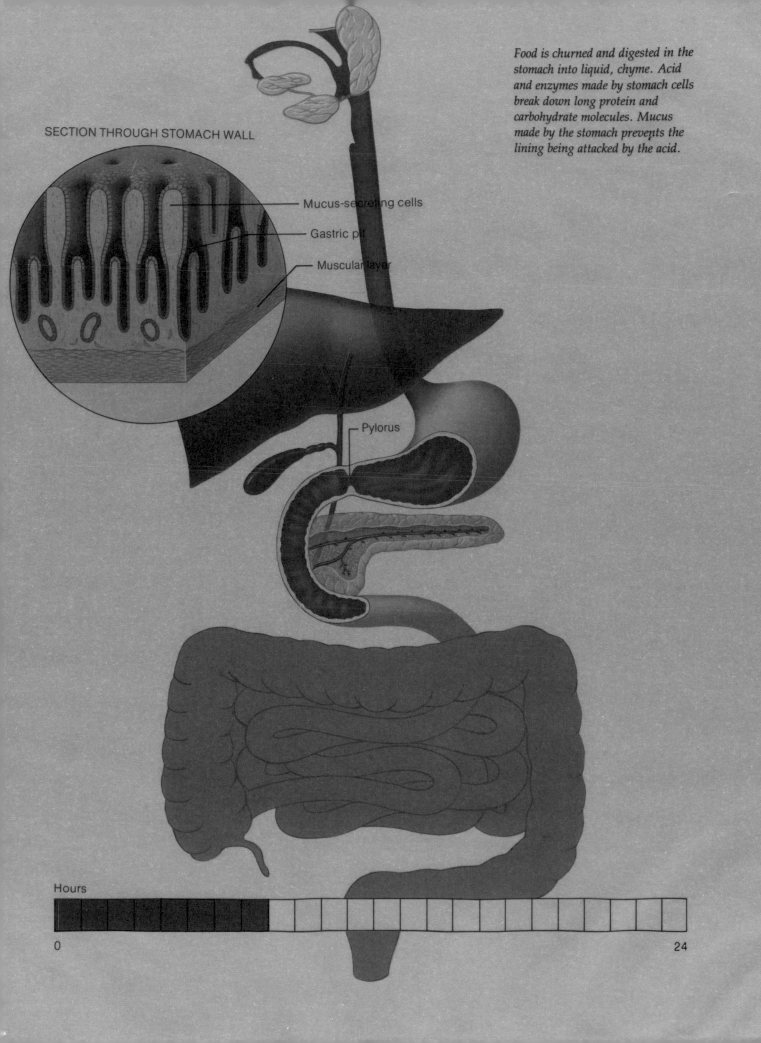

SECTION THROUGH STOMACH WALL

Food is churned and digested in the
stomach into liquid, chyme. Acid
and enzymes made by stomach cells
break down long protein and
carbohydrate molecules. Mucus
made by the stomach prevents the
lining being attacked by the acid.

Mucus-secreting cells

Gastric pit

Muscular layer

Pylorus

Hours

0

24

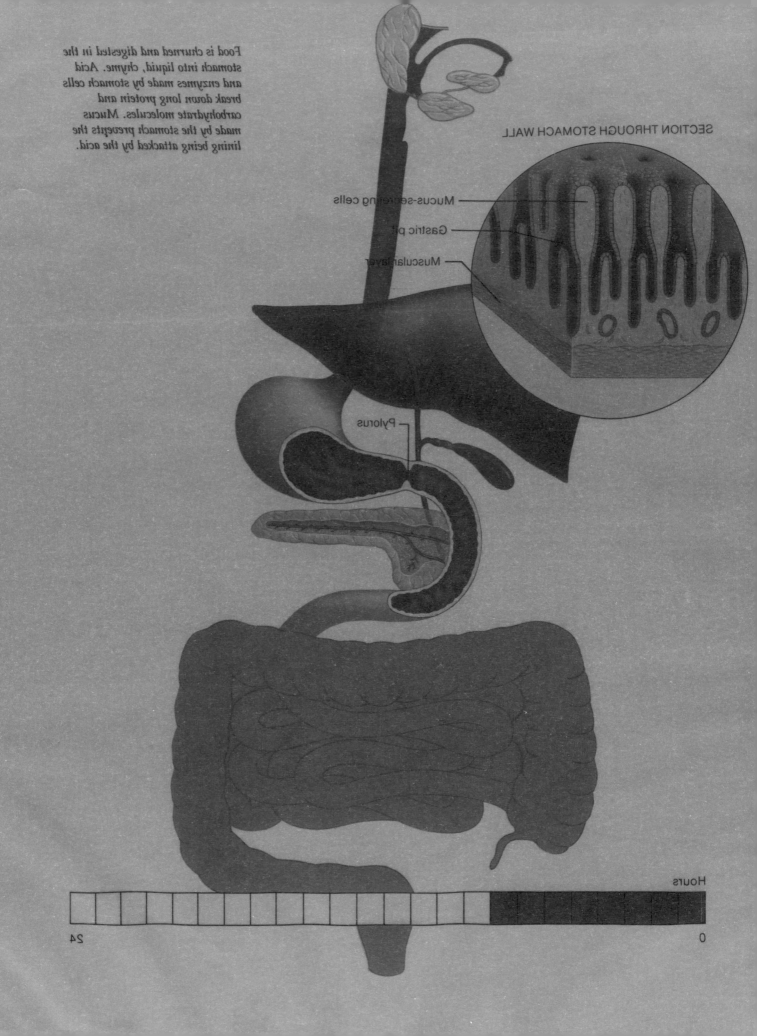

Hours

24

0

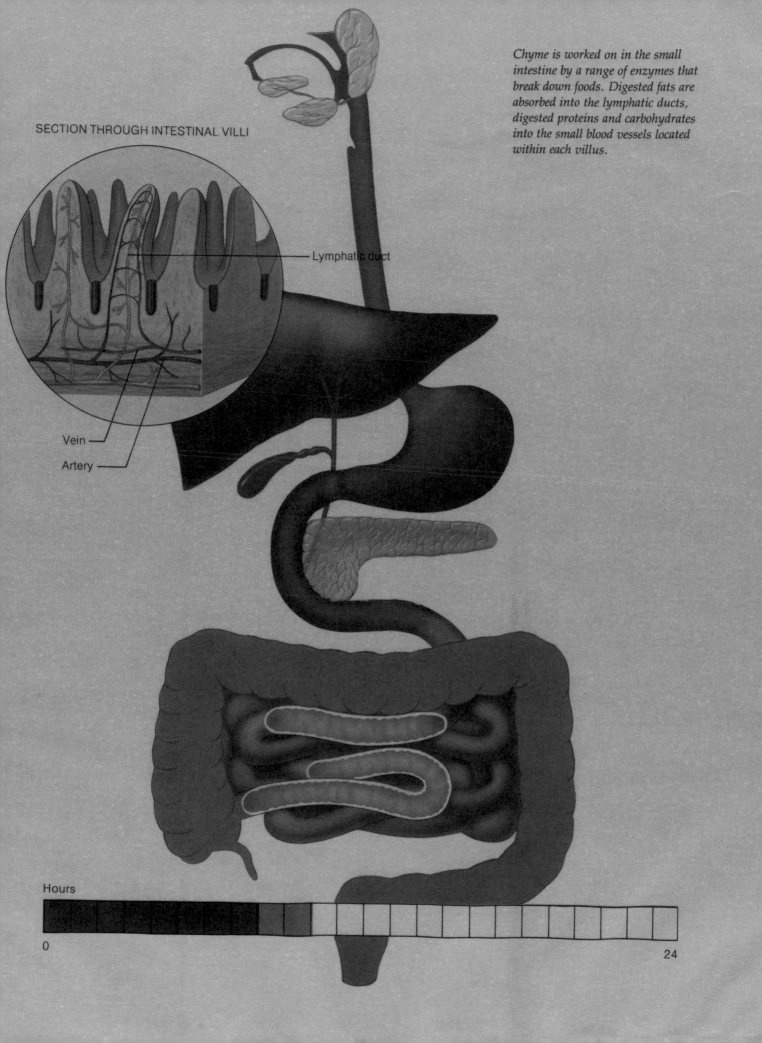

SECTION THROUGH INTESTINAL VILLI

Chyme is worked on in the small
intestine by a range of enzymes that
break down foods. Digested fats are
absorbed into the lymphatic ducts,
digested proteins and carbohydrates
into the small blood vessels located
within each villus.

Lymphatic duct

Vein

Artery

Hours

0

24

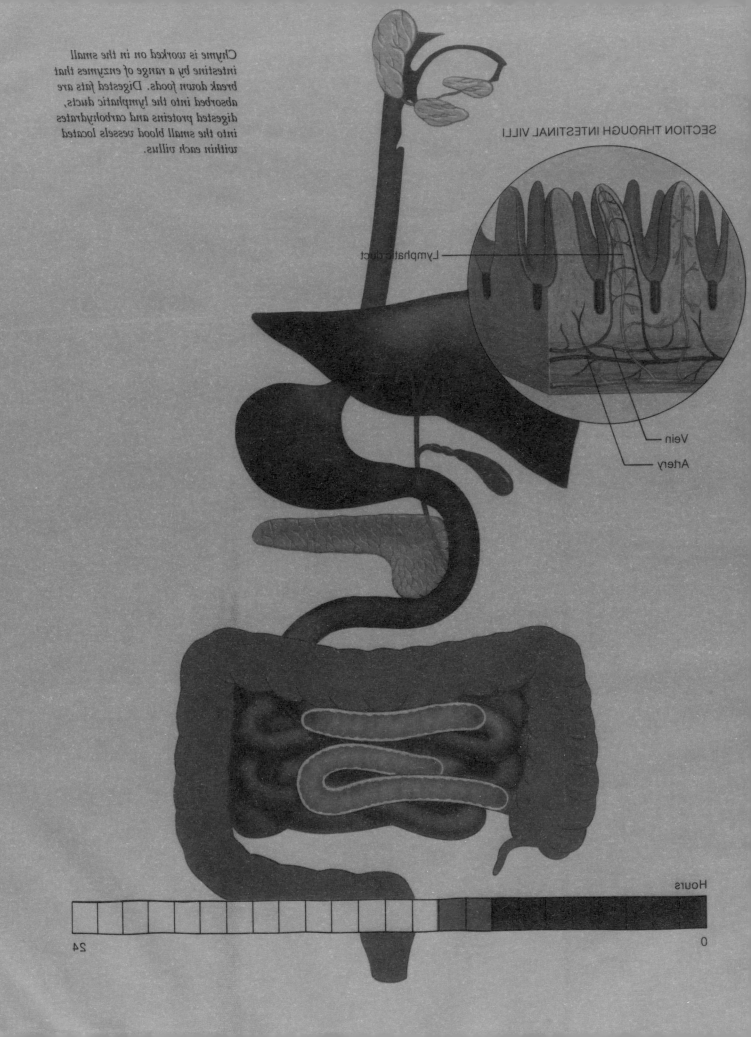

Hours

24

0

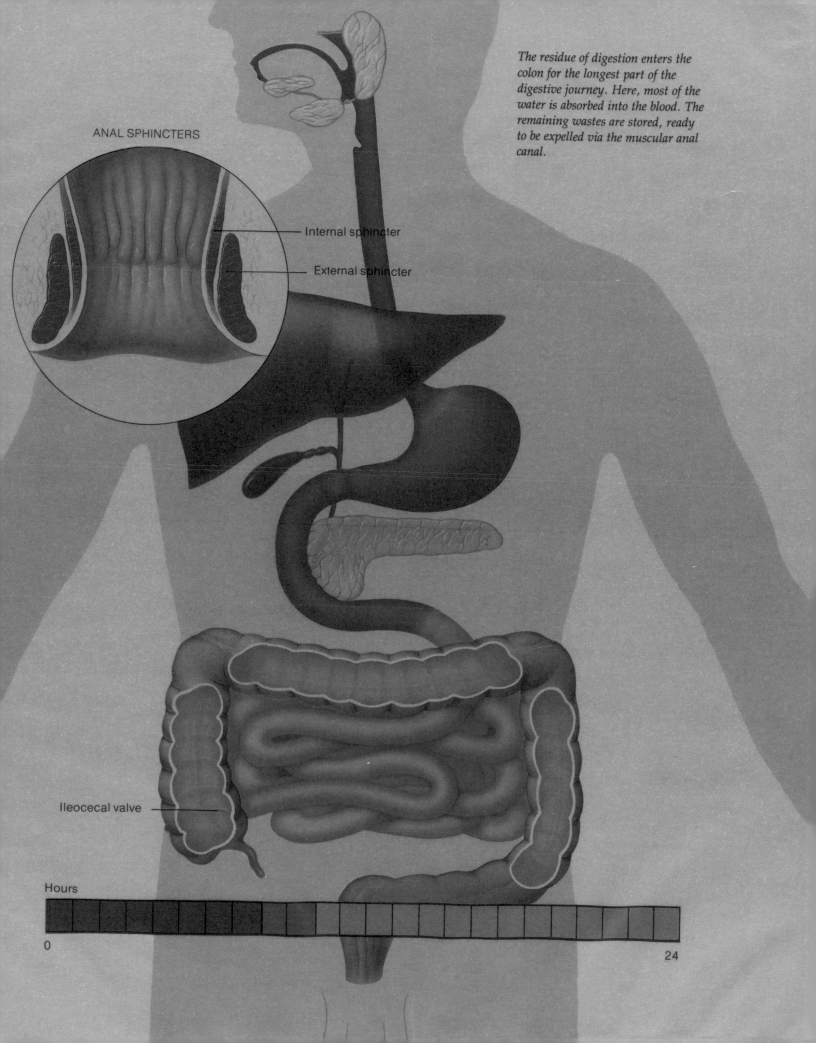

ANAL SPHINCTERS

Internal sphincter

External sphincter

Ileocecal valve

The residue of digestion enters the colon for the longest part of the digestive journey. Here, most of the water is absorbed into the blood. The remaining wastes are stored, ready to be expelled via the muscular anal canal.

Hours

0

24

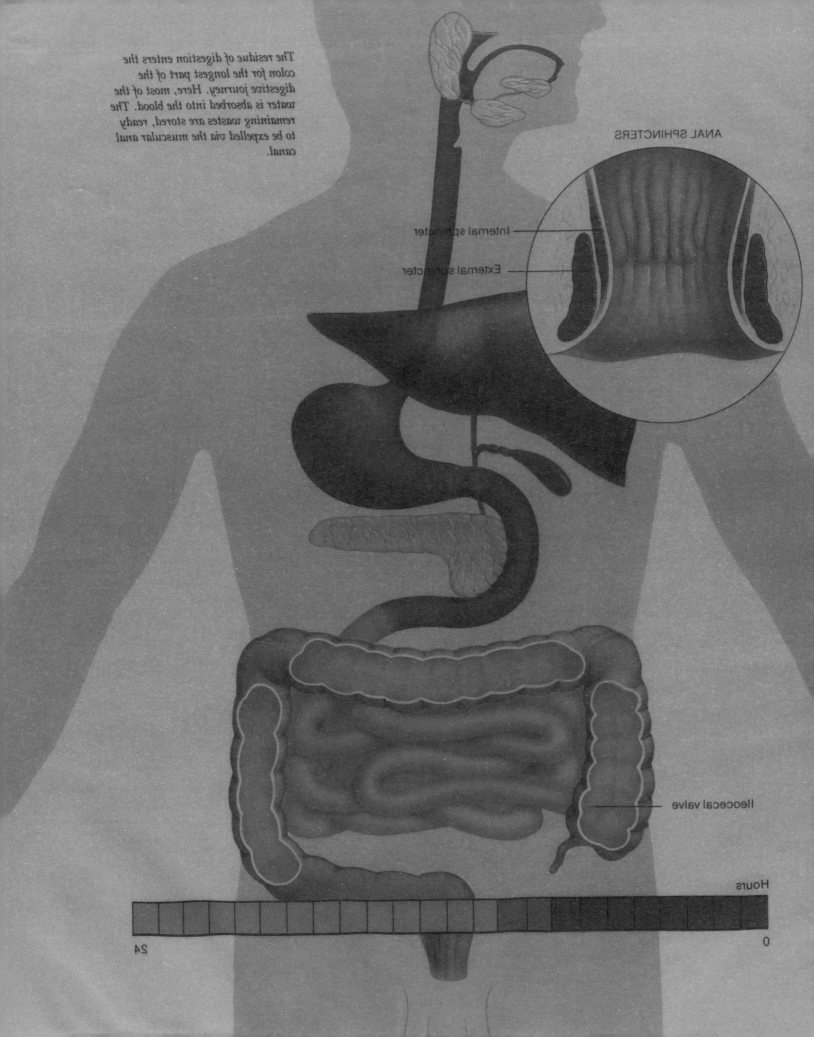

External sphincter

Ileocecal valve

Hours

24   0

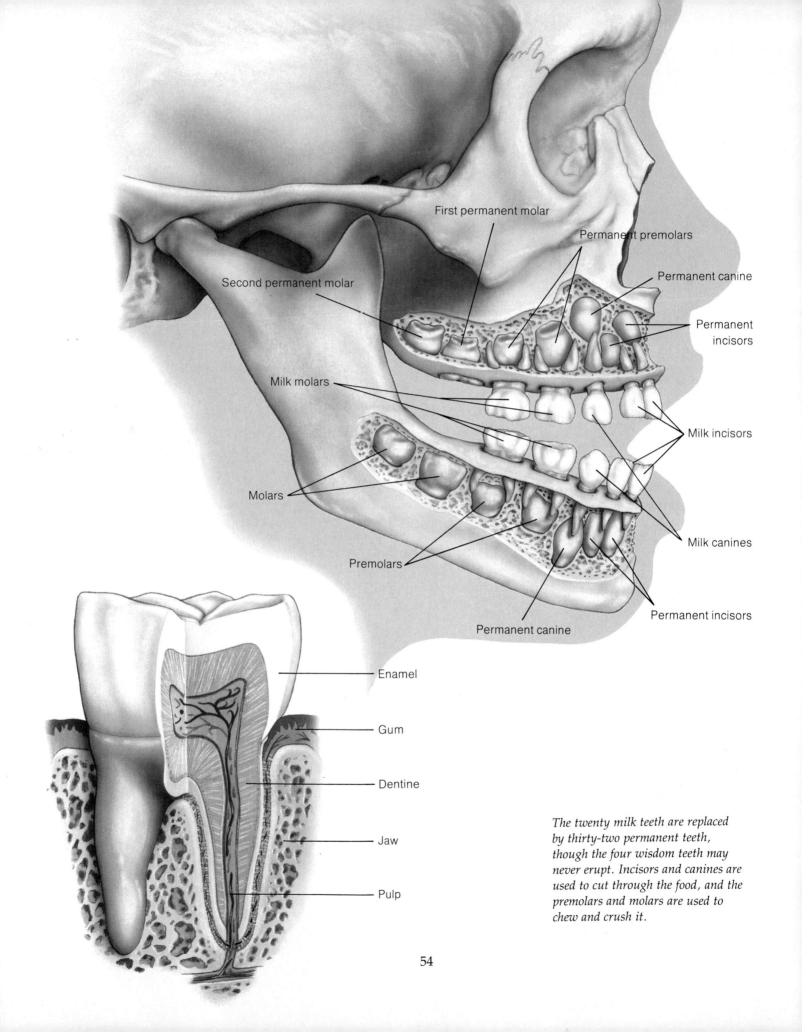

First permanent molar

Permanent premolars

Permanent canine

Permanent incisors

Second permanent molar

Milk molars

Milk incisors

Molars

Premolars

Milk canines

Permanent incisors

Permanent canine

Enamel

Gum

Dentine

Jaw

Pulp

The twenty milk teeth are replaced by thirty-two permanent teeth, though the four wisdom teeth may never erupt. Incisors and canines are used to cut through the food, and the premolars and molars are used to chew and crush it.

54

Theodore Billroth

The Great Pioneer

"Criticism is the principal need of our day, and for this, knowledge, experience and calm are requisite." These words were spoken by Theodore Billroth, one of the greatest surgeons who ever lived.

Born on April 26, 1829, at Bergen, on the island of Rügen, Billroth was of Swedish parentage. As a young man, he was an accomplished musician and wished to make a career in music but was encouraged by his mother — his father died when Billroth was five — to take up the more reliable and lucrative career of medicine. He qualified as a doctor in 1852, having studied at Greifswald, Göttingen and Berlin. After qualifying, he spent some time touring the principal medical schools of Europe before accepting a post as assistant to Bernard Langenbeck in Berlin. This appointment was to start Billroth on his career. At this time, surgery as a specialist branch of medicine was experiencing a rebirth, thanks to the invention of general anesthesia. Operations could now be performed on internal organs, such as the stomach and the intestines, and this opened up a whole new field of preoperative diagnosis.

In 1860, Billroth was appointed as professor of surgery in Zürich. Here he had a chance to develop

his theories of self-criticism, and during his seven years there he published many reports, listing his failures as well as his successes. One of his most famous reports is on the cause of wound infection. He was among the first to point out the usefulness of regular temperature taking as an indicator of the course of an infective process in surgery. He also published a monumental work on surgical pathology and treatment, *Die Allgemeine chirurgische Pathologie und Therapie.*

Billroth moved to the University of Vienna as a professor of surgery in 1867, and it was in this city that he spent the rest of his working life. As well as writing and teaching, Billroth became famous as a surgeon, thanks to his skill at,

and his development of, surgery of the stomach. In 1881, a female patient suffering from an internal obstruction was found on operation to have stomach cancer. Billroth removed the tumour along with part of the stomach and joined the remainder of the stomach to the duodenum with sutures. The continuity of the gastrointestinal tract was restored, the operation was successful, and the patient survived. Unfortunately, she died a few weeks later from a secondary cancer, but the operation was crucial to the development of abdominal surgery.

Billroth also founded the modern concept of writing up detailed reports on operations carried out within a department, noting operative mortality and complications, and he instigated a five-year follow up of these cases. However, although he thought these records necessary, he once wrote; "Statistics are like women, mirrors of purest virtue and truth, or like whores, to use as one pleases."

Throughout his life, Billroth continued to be an accomplished and enthusiastic musician. He was a close friend of Brahms, who often used to play his works to Billroth before they were published. He died in Abbazia on the Adriatic on February 6, 1894.

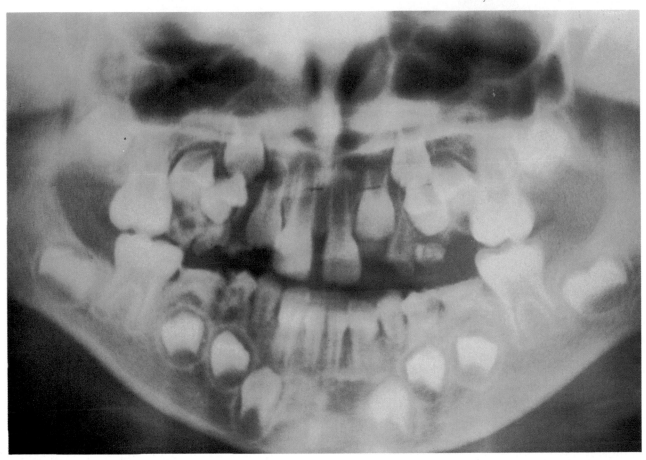

Because the permanent teeth develop from early on in fetal life, poor diet and disease in the mother at this time can have serious and even permanent effects on the teeth of her children. In particular, adequate supplies of vitamin D are essential for the proper growth of the dentine body of the tooth and vitamin A for the normal coating of the dentine with enamel. Similarly, infections in the mother, such as syphilis and rubella, can seriously damage the teeth long before they erupt.

The first teeth appear in the jaw from the seventh to the twenty-fourth month of life, starting usually with the front, cutting incisor teeth and ending with the rear, flat-topped grinding molars. The eruption into the gums of an infant's first teeth has, in the past, been blamed for many of the illnesses and discomforts that befall children at this age.

The process of eruption of the permanent teeth starts at the age of about seven years and continues into an individual's twenties. First the milk teeth must be shed, producing the familiar gap-toothed grin of midchildhood. Since there are more permanent teeth than milk teeth, some permanent teeth must erupt through the gums without following a fallen deciduous tooth. Usually this occurs without discomfort; the so-called "wisdom teeth," however, which may not erupt until late teens or early twenties, can cause problems.

Wisdom teeth are the last molar teeth, situated at the back of the jaw. Often, owing to the shape of the jaw, they erupt at an angle and are impacted against the teeth in front. The flap of gum covering the incompletely erupted tooth allows debris, such as food particles, to collect under it, resulting in infection and the danger of abscess formation. This may produce spasm of the muscles of chewing, called "trismus," when the mouth cannot be opened fully and sometimes not at all.

Anchored to the floor of the mouth and slung at the rear from muscles attached to a spiky outgrowth at the base of the skull is the tongue, which itself is a powerful muscle. Covering this is the specialized membrane, known as the lingual membrane, on part of which are set the taste buds, the organs of taste and flavor. The muscle fibers of the tongue are richly supplied with nerves, enabling the tongue to move food around in the mouth; they also control the exit of air from the mouth, making speech possible. Carefully coordinated movements of the tongue are necessary to commence the task of swallowing food.

Disease in the Mouth

Because the mouth is so accessible, disease is easily seen as long as it occurs near the front. Doctors can learn a great deal just by looking at the gums, for instance. Here may be the telltale mark of poisoning with substances such as lead or bismuth (used to treat syphilis before the introduction of penicillin after World War II), for metallic poisons leave a line on the gums near the teeth. Painters using lead-based paints, and workers in the lead industry who are inadequately protected, may have a series of gray-black dots visible just below the boundary of tooth and gum. A similar line indicates mercury poisoning. This can result from industrial spillage of liquid mercury which easily atomizes in the air and is then inhaled.

Cancerous growths in the mouth make up one in twenty of all human cancers. Malignant growths on the lips are a frequent finding in areas of the world where strong sunlight is usual. The tongue is another site of cancer, often starting as an innocuous looking area of whitening of the normally pink surface. A frequent cause of tongue and mouth cancer is tobacco smoking.

Treatment of tongue and mouth cancers is often successful because of their frequent early detection by people inspecting their mouths in the mirror. Surgical removal of even quite large portions of the tongue's surface can be achieved without serious loss of its function. Radiation therapy often supplements this surgery.

In addition to detecting these disorders in the mouth, physicians take a keen interest in any color changes that occur there. For example, patches of blue-brown pigmentation appear in the mouths of individuals suffering from a deficiency of adrenal hormones in Addison's disease. Less commonly, those with a curious inherited condition of multiple, fleshy, polypous outgrowths in their small bowel may be identified by frecklelike patches of pigmentation in and around the lips and gums.

Examination of the tongue in the mirror is a daily habit in many individuals, but most of the changes visible signal no harm. Rather they serve to alarm the possessor of the tongue. On occasion, the papillae, small fleshy outgrowths on the tongue lining, may grow to unusual size and can often become discolored, particularly if the individual is not careful with oral hygiene. This results in the so-called "black hairy tongue," an entirely benign condition. Of more serious significance is the tongue which increases in size. This may be the sign of a more complex disease, known as amyloidosis,

where curious starchy deposits are laid down in many tissues as a result of a disorder of the body's immune defense system.

In certain stages of scarlet fever the upper layers of the tongue may be stripped off. This stripping is caused by a toxin, or poison, released from the infecting bacteria and results in the appropriately named "strawberry tongue." Yet another tongue symptom occurs in various types of anemia; here the tongue may become excessively smooth.

The Glands of the Mouth

The sensitive lining of the mouth must be kept moist at all times, so small glands scattered through the mouth constantly produce small quantities of saliva. However, when food is eaten, larger amounts of saliva, rich in the starch-breaking enzyme amylase, are needed to lubricate both the process of chewing and the subsequent swallowing of the meal. To cope with this, there are three more specialized salivary glands on each side of the mouth that secrete saliva when stimulated. These

glands are the parotid (a large structure resting over the angle of the jaw near the ear), the submandibular (tucked into the lower jaw) and the small, almond shaped sublingual glands under the tongue. Saliva production is stimulated by nerve endings, whose actions are coordinated by the brainstem centers that respond to the presence, or even the possibility, of food in the mouth.

The most frequent problem with the salivary glands is the virus infection of mumps. In this, the glands, particularly the parotid, swell up and become painful, producing the chipmunk face of the sufferer; sometimes all the salivary glands may swell one after another or simultaneously and the swelling causes difficulty in swallowing and talking. Although children usually recover fairly quickly, adults — particularly males — may be in danger if they contract mumps. The infection may spread through the blood to other organs, including the testicles, the pancreas and the lining membranes of the brain, where a benign meningitis may occur. When a man's testicles are affected, the sufferer

may be rendered sterile.

The parotid gland can be the site of malignant tumors, whose removal is often made surgically complex since the nerve to the muscles of the face runs through it and can easily be damaged.

Other problems which may arise in the salivary glands include the formation of stones in the long ducts leading from the glands to the mouth, blocking the passage of the saliva and causing painful distension of the glands when the start of a meal activates the glands to pour out saliva. Other diseases can lead to an uncomfortably dry mouth, or xerostomia. Sometimes this is the result of an attack on the glands by the body's own immune system. In other circumstances it may be the result of side effects from drugs, especially certain types of anti-depressants.

Since the mouth is in constant use for social purposes — including kissing — social propriety demands that the breath smell sweet. Halitosis, or bad breath, may be the result of serious disease in the mouth or chest but more often results from poor oral hygiene and lack of attention to gum infections. Habits such as smoking contribute to the bad breath of many individuals. However, physicians can gain diagnostic clues about more serious generalized disease of the liver or kidneys and the presence of diabetic acidosis in unconscious patients simply by smelling the breath.

At the back of the mouth, the palate, or root of the mouth, ends with two flaplike pillars, behind which are the tonsils; these are obvious in children but shrivel to almost nothing in the adult. These fleshy outgrowths of lymphatic tissue from the side walls of the rear of the mouth consist mainly of collections of immune defense cells and mark one line of defense against invading infections. Often removed unnecessarily, after being blamed for various minor ills in children or along with infected, enlarged adenoids, the tonsils with their sister glands, the adenoids farther back in the throat, may become infected by bacteria and viruses. Usually this is a simple problem, quickly responding to antibiotic treatment if a bacterial in-

By the time food is ready to be swallowed, the original mouthful has been transformed into a soft ball. The flexible tongue pushes the bolus up against the roof of the mouth and into the pharynx. As soon as the food reaches the pharynx, the soft palate is pushed upward by the tongue to shut off the inner entrance to the nose, and the epiglottis snaps down over the entrance to the tubes that lead to the lungs. The food then passes into the esophagus. Once in the esophagus, waves of muscle contractions and relaxations — peristalsis — move the bolus of food down to the entrance of the stomach.

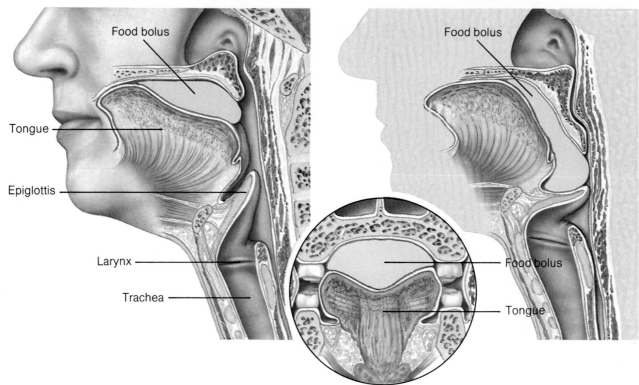

CROSS SECTION

fection is present. Frequent attacks of tonsillitis may indicate that a tonsillectomy would help, but physicians are now more wary of recommending this form of surgery.

Swallowing

At the start of a swallow, the tongue tip touches the roof of the mouth at the hard, front part of the palate. Trapping the ball of masticated food behind it, the body of the tongue then rises from the front, pushing the food toward the waiting pharynx at the back of the throat. The pharynx is an apparatus of muscles and specialized tissues used for the accurate transport of food to the gullet and from there to the stomach. At the back of the pharynx is a flap of tissue, the epiglottis, which closes down over the top of the windpipe as each mouthful of food is swallowed. This protective mechanism, controlled by reflex action, ensures against the risk of asphyxiation due to blockage of the windpipe.

As the food reaches the pharynx, the soft rear part of the palate is lifted up, blocking the nasal passages and preventing regurgitation of the now fairly liquid ball of food, called a bolus, back through the nose. At about the same time, the vocal cords across the top of the voice box (larynx) shut and the larynx is raised up against the epiglottis. As this occurs, a wave of contractions in the muscles surrounding the pharynx forces the partly digested food back into the top of the gullet, which lies behind the windpipe.

Once it is in the gullet, or esophagus, successive waves of contractions move the bolus of food down to the entrance of the stomach at the top of the abdomen in a process called peristalisis. The upper one-third of the muscles of the esophagus are under voluntary control, the middle third are a mixture of this voluntarily controlled muscle and automatically controlled smooth muscle, while the lower third is smooth muscle alone. Thus, once the process of swallowing has started, it rapidly becomes automatic. While gravity helps the downward passage of food to the stomach, it is not necessary to be upright when eating, and if you wanted to, you could eat a three course meal while standing on your head.

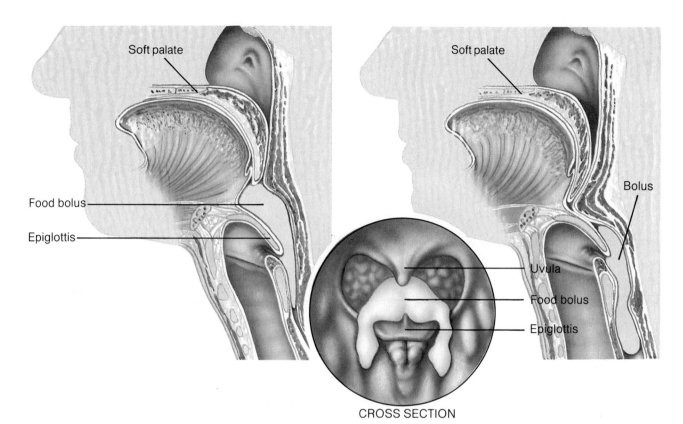

Soft palate

Food bolus

Epiglottis

Soft palate

Bolus

Uvula

Food bolus

Epiglottis

CROSS SECTION

The air passages are also protected by their extreme sensitivity to the presence of anything solid within them. The fits of choking and coughing that follow food going down the "wrong way" are an uncomfortable experience and testimony to the vigor of this defense. On occasions, the reaction is so extreme that the vocal cords shut tight, preventing breathing. While the traditional cure for such bouts of choking has always been a sharp slap on the back, modern research has shown that this is relatively ineffective.

In the mid 1970s the American physician Henry Jay Heimlich advocated an anatomically accurate cure for severe choking attacks which can be life saving. In the Heimlich maneuver, a sharp, unexpected push is applied to the abdomen of the sufferer so that a sudden rush of air is pushed out of the chest, propelling misdirected food out of the windpipe back into the mouth. Many people die each year from choking — the so-called "cafe coronary," though it has no relation to coronary thrombosis — and it is likely that the widespread application of the Heimlich maneuver to choking people in place of the traditional practice of thumping them on their backs would save lives.

Peristalsis

The whole of the gut, from the top of the esophagus to the rectum, is structured to make a snakelike movement, whereby its contents are progressively pushed through the digestive system. The principal components of this action are the same throughout the gut, though the speed and force of the contractions vary from one area to another. In the gut, peristalsis is effected by muscle layers of varying thickness and strength. On the outside there are muscle fibers arranged in line with the length of the gut tube; inside this, there is a layer of muscles whose fibers run at right angles to the outer layer, thus encircling the gut. Between these two layers there is the "myenteric" plexus of delicate nerve fibers. Connected to this network of nerve fibers is another plexus, just below the inner mucous membrane lining of the gut, the submucus plexus. The stretching of this inner layer of nerve fibers by food or gas in the bowel starts a reflex

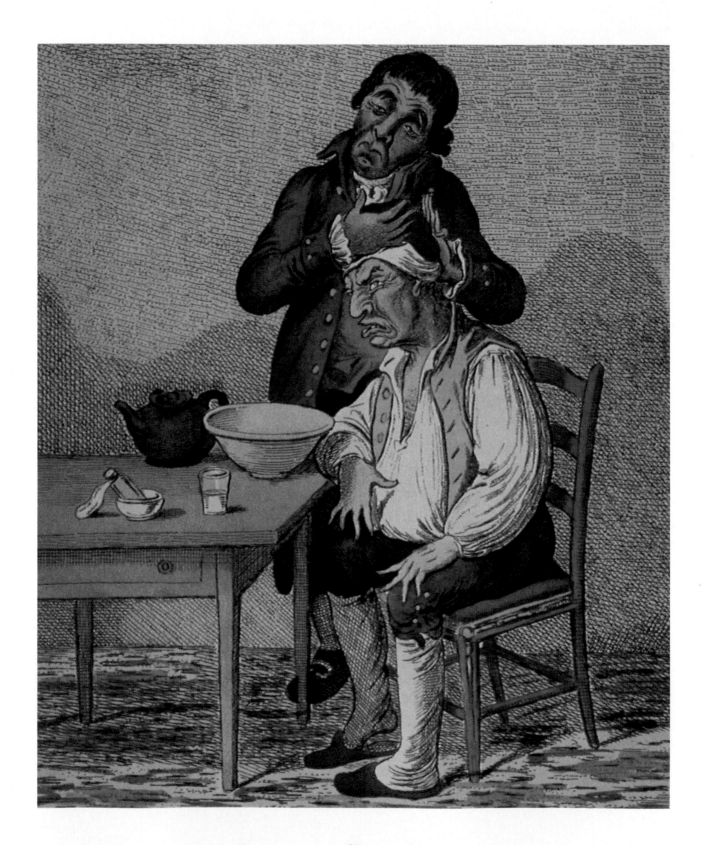

62

The unfortunate patient (left) has taken a gentle emetic to make him sick. Vomiting involves reverse peristalsis — muscle contractions cause the stomach contents to be forcibly ejected.

A good view of the stomach can be vital to correct diagnosis. In the nineteenth century this was possible, but patients had to undergo an uncomfortable procedure involving the insertion of rigid tubes (below).

The invention of the fiberoptic gastroscope, or endoscope (bottom), has enabled physicians not only to see into the upper gut but also to perform minor operations with little discomfort to the patient.

contraction of the muscle immediately behind the food, which causes the circular muscle to contract and the muscle in front to relax. This, in turn, causes the food to be pushed down the gut and starts the process in the next segment. Once started, a wave of contractions passes down the gut behind a wave of relaxations.

The esophagus stretches in three smooth curves from the back of the pharynx to just below the diaphragm, where it meets the stomach at an angle. At the bottom of the esophagus is a ring of muscle, or sphincter, which tightly refuses the contents of the stomach entry into the gullet. It is important to protect the lower end of the esophagus from the acidic contents of the stomach, which would otherwise irritate the gullet's sensitive lining. The food, whose journey from mouth to stomach started with the swallow, travels by peristalsis down the esophagus, lubricated by mucus secreted by small glands in its lining. At the gastro-esophageal junction, the ringlike sphincter relaxes when a wave of peristalsis reaches it and allows the bolus to drop into the stomach. At this point, the slightly alkaline food meets the gastric juices, which are acid, and digestion starts in earnest.

Esophageal Problems

A common symptom of problems at the junction of the esophagus and stomach is acid heartburn. This has nothing to do with the heart and is felt as a burning pain at the bottom of the chest which rises to the throat. Its cause is failure of the sphincter of the lower gullet to stop acid refluxing into the chest from the stomach. Often this is caused by herniation, or bulging, of the top of the stomach into the chest so that the muscle at the bottom of the gullet is stretched and cannot close properly — the so-called hiatus hernia.

Heartburn is most often felt on lying down or when the stomach is full. This situation arises especially in extremely obese people, when the pressure from the abdominal contents tends to push the top of the stomach up through the opening in the diaphragm. The acid escaping into the lower esophagus causes an irritation and inflammation of the lining and this causes pain. If the condition persists for some time, ulcers will form at the bottom of the esophagus, worsening the pain.

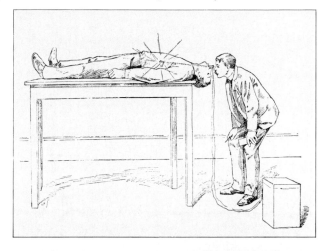

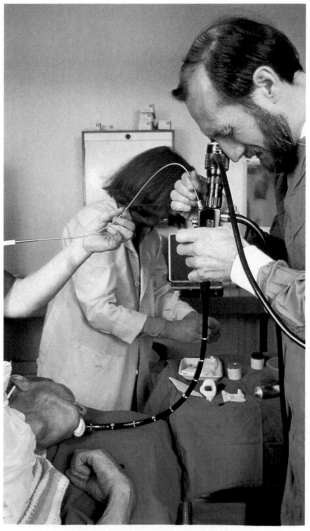

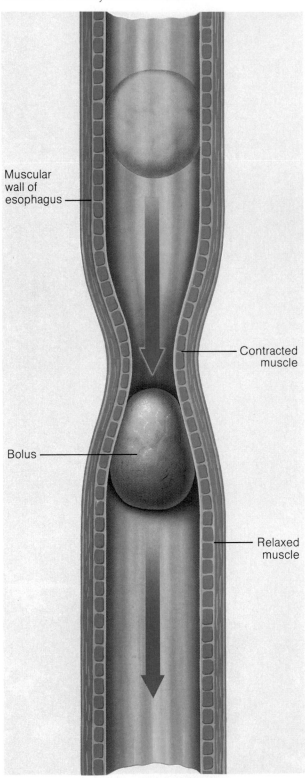

Peristalsis is the mechanism whereby food is forced along the alimentary canal by a series of involuntary muscular contractions and relaxations. The speed of the contractions varies from area to area.

Muscular wall of esophagus

Contracted muscle

Bolus

Relaxed muscle

Sometimes the symptom can be relieved by the following combination of measures: losing weight; sleeping sitting up; and eating small, frequent meals, possibly aided by antacid medication. Surgery may correct the displacement of the upper part of the stomach if these measures fail.

The most common symptom of disease in the esophagus is difficulty in swallowing, called dyspagagia. The food seems to stick in the chest and a second swallow may be necessary to accomplish the delivery of food to the stomach. The most sinister cause of this difficulty is a tumor in the lining membrane of the esophagus. Although in North America and Europe this tumor accounts for fewer than one in ten of all cancers, in parts of the Far East it is the commonest cancer found.

It is believed that dietary factors are important to the appearance of this malignant growth of the cells of the inner part of the gullet, but what factor is involved is still uncertain. Surgery is needed to remove the growth or, failing this, a temporary improvement in swallowing can be achieved by inserting tubes to pass through the obstructing growth. When the tumor has not spread outside the gullet, the part containing the growth can be removed and replaced by pieces from the patient's large bowel.

Pain and difficulty in swallowing may occur if certain nerves between the muscle layers of the esophagus are damaged and the gullet goes into spasm on swallowing. This seems to occur particularly in older individuals and can cause a pain so severe that it may mimic the pain of coronary thrombosis. Less commonly, the rare disease systemic sclerosis can cause swallowing difficulties. In this disease, the gullet is one of many tissues, including the skin and other parts of the bowel, which become progressively invaded by fibrous tissue instead of normal elastic connecting tissues.

When certain nerves controlling peristalsis at the bottom of the esophagus are damaged, a condition called "achalasia" of the esophagus develops. Here, the bottom of the gullet refuses to open. Swallowed foods and liquids build up in the esophagus above the obstruction, and the tube becomes dilated and bloated with rotting food — a potent cause of bad breath. The condition is dangerous when this putrid food spills over into the air

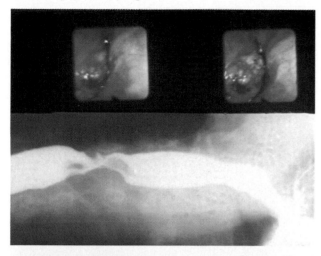

passages to cause pneumonia and fibrosis of the lungs. Since food is no longer entering the stomach in normal quantities, the nutrition of the patient suffers accordingly. Treatment is by surgical widening of the muscular bottom of the gullet, sometimes by slitting some of its muscle fibers, but often by forceful stretching of the muscle ring from above. Some drugs, such as the smooth muscle relaxant amyl nitrate, may be used to open the sphincter while surgery is awaited.

Infections of the esophagus are rare but do occur, particularly in individuals whose resistance to infection is reduced, for example by the drugs used in heart and kidney transplants to prevent rejection. These may cause the yeast candida to invade the gullet, producing esophageal thrush. This is a very unpleasant disease, causing constant swallowing difficulty and severely reducing the sufferer's appetite for food as a result. The herpes virus can sometimes spread into the gullet from the mouth, where it more usually causes cold sores. The virus causes a severe and painful esophagitis, though new antiviral drugs offer promise of an early cure for victims of this disease.

The Stomach

The stomach plays two main parts in digestion. The upper baglike portion, the fundus of the stomach, acts as a hopper to hold the food delivered from the esophagus and to feed it to the lower two-thirds. The other portion, the antrum, is surrounded by smooth muscles which churn the food, fermenting the solids and mixing them with protein-splitting enzymes. The average adult stomach holds three pints and manufactures the same volume of gastric juice in each twenty-four hour period. The stomach contains hydrochloric acid, manufactured by special parietal cells in the gastric mucous membrane. The acid solution is formed by the active metabolism of the parietal cells and is pushed out into the stomach. This transforms the mildly alkaline chewed food by helping to break down the large pieces of which it is constituted into much smaller ones. This process is aided by an enzyme called pepsin made by other cells in the gastric lining.

The control of the acid secretion of the stomach is of great importance since it must be coordinated with the appearance of the food on which it is to work. Its presence in large quantities without the neutralizing effect of food might lead to damage to the stomach wall itself, as can occur in cases of peptic ulceration.

The Lining of the Stomach

In keeping with its several functions, the lining mucous membrane of the stomach is of three distinct types. At the top of the stomach, in the cardia, the lining contains simple mucus-secreting cells, which bathe the stomach in a thin protective film and help lubricate the churning function of the lower parts. The mucous membrane of the main part, or body, of the stomach, the parietal mucosa, contains the acid-secreting cells already described. At the far end of the stomach, near the junction with the duodenum, are more simple secreting glands and those that make the hormone gastrin.

The stomach is well supplied with nerves, most of which come from the autonomic nervous system — that part of the nervous system that controls the automatic or involuntary aspects of the body's activities. The most important nerve is the vagus nerve which arises from centers in the medulla, at the head of the spinal cord, and which passes down to inervate many structures within the abdominal cavity. However, the nervous system is not the sole mechanism which carries messages to the stomach from elsewhere; hormones are involved as well. Hormones are chemical messengers which are secreted in one part of the body and carried in the blood stream to produce an effect elsewhere.

The most important element in the control of gastric emptying is the enterogastric reflex. As the duodenum and intestine become filled with food from the stomach, they initiate the reflex, which is carried to the stomach via the vagal nerve and by other small nerves that are part of the celiac plexus. The effect of the reflex is to reduce the amount of movement in the stomach wall and hence to slow down the rate of emptying into the duodenum. The arrival of food in the duodenum and the degree of acidity that results are sufficient to trigger the reflex. The hormones secretin and cholecystokinin are also released from the mucosa of the duodenum. They are released in response to the arrival of food in the duodenum and have the

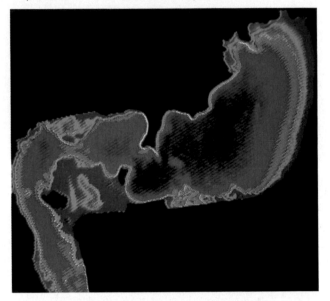

effect of limiting both gastric motility and the secretion of acid by the stomach wall. These two hormones are also important in triggering other effects within the digestive system, especially the secretion of digestive substances by the gallbladder and the pancreas.

Digestion by the Stomach

The stout muscle coats of the body of the stomach churn food for some hours, while the potent gastric juice starts the process of digestion by chopping up the long protein and polysaccharide molecules of

This fiberoptic view of a patient's stomach shows the light discoloration of a gastric ulcer. An ulcer such as this might eat a hole in the stomach wall or it might possibly disappear spontaneously.

The ulcer in the picture (below center) is a duodenal ulcer. This type is two or three times as common as a gastric, or stomach, ulcer. A gastric ulcer in the process of healing is shown in the bottom

picture. Although they do not seem to be closely associated with over-production of stomach acid, treatment with antacid medicines or drugs that reduce acid secretion will cure most ulcers.

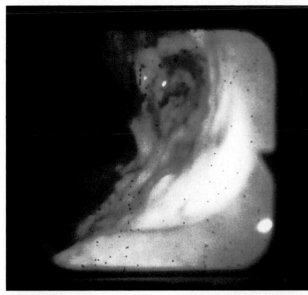

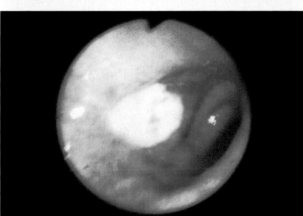

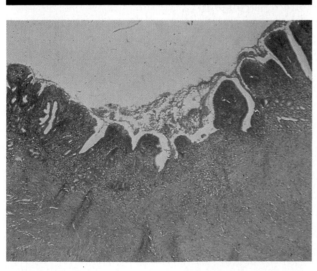

the undigested food into small pieces. The result is a sticky liquid called chyme. This is released into the duodenum through the pylorus, the muscular ring at the far end of the stomach where the small bowel starts. As squirts of acid chyme enter the duodenum, gastric emptying is spurred on by the release of hormones by the top end of the small bowel which also signal to the stomach to halve acid production.

While much of the mouth can be easily viewed by the physician in search of disease, the esophagus and stomach have, until recent decades, remained hidden from view. Advances in technology have now brought the upper gut into full view. Originally rigid tubes with clumsy lighting apparatus were inserted into the gullet and stomach in a type of medical sword-swallowing act.

Modern medical technology can direct light around corners with the aid of fine quartz fibers. When these fibers are perfectly aligned in a long flexible tube, the inside of the upper gut can be viewed in full color without the need for anesthesia and with minimal discomfort to the ill individual. This has revolutionized the treatment and diagnosis of all gut diseases but nowhere so completely as in the upper gut. Such conditions as peptic ulcers can now be seen while previously their presence was diagnosed indirectly with X rays. Bleeding points in the esophagus and stomach can be identified with these instruments. Fine probes passed down side ports in the instrument allow air to be pushed into the stomach, or blood obscuring the view to be wafted away with puffs of carbon dioxide.

Even more recent technological developments have included the perfection of argon lasers with which bleeding arteries can be heat sealed or cauterized to halt hemorrhages that might otherwise have necessitated major abdominal surgery. Tiny forceps can also be passed down the instrument to take pieces of tissue from suspected areas. These are examined microscopically for the presence of cancers at an early and treatable stage. No one instrument can be said to have changed the face of gastroenterology more in the last twenty years than the fiberoptic gastroscope, which is now as vital to the gut specialist as the stethoscope is to other physicians.

With a duodenal ulcer (below), acid leaving the stomach erodes the duodenum's lining. As the ulcer crater gets deeper, there is the risk of leakage into the abdominal cavity which could result in peritonitis.

When a gastric ulcer is formed (below), the stomach acid attacks the stomach wall itself. Causes are either overproduction of acid or a failure of the mucosal protection provided by the stomach's lining.

In the condition known as esophageal reflux (below), stomach acid washes back into the esophagus and causes irritation of the tissues lining the esophagus. This condition is common during pregnancy.

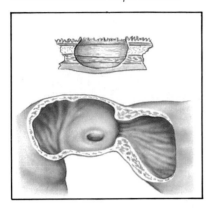

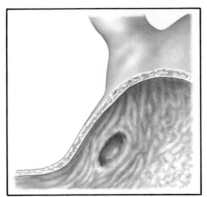

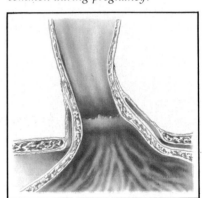

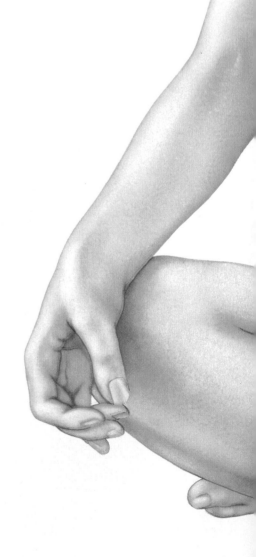

Ulcers

The stomach lining lies in a bath of acid of its own making, and it needs a defense mechanism to protect its cells from being digested. This comes mainly from the protective mucous film that coats its inner walls. Breakdown in this mucosal defense produces ulceration of the stomach lining. This is an event which is responsible for the death of 10,000 Americans every year. The causes are a combination of overproduction of acid with an associated failure of the mucosal barrier.

Ulcers appear at the junctions of the different types of gastric lining. So, in the stomach, gastric ulcers occur at the meeting of the mucus-secreting cardia and the gastrin-secreting antrum, with the main acid-secreting body of the stomach in between. In the duodenum, ulcers are usually restricted to the first half inch beyond the muscular gates of the pylorus. In gastric ulcers, the cause seems to be mainly a breakdown in the mucosal resistance, and men and women are affected in about equal proportion. In duodenal ulcers, the two factors are combined: in some sufferers the acid production is normal, but the lining of the stomach is not capable of dealing with this normal acid challenge. In about one-third of duodenal ulcer victims, a majority of whom are men, the acid production is unusually high, and the normal defenses are insufficient to prevent ulceration.

Many factors have been found to provoke ulcers, but the causes often remain mysterious. As with many diseases which plague modern man, psychological stress has been cited as a cause, but modern gastroenterologists view this with increasing scepticism. There is, however, a genetic

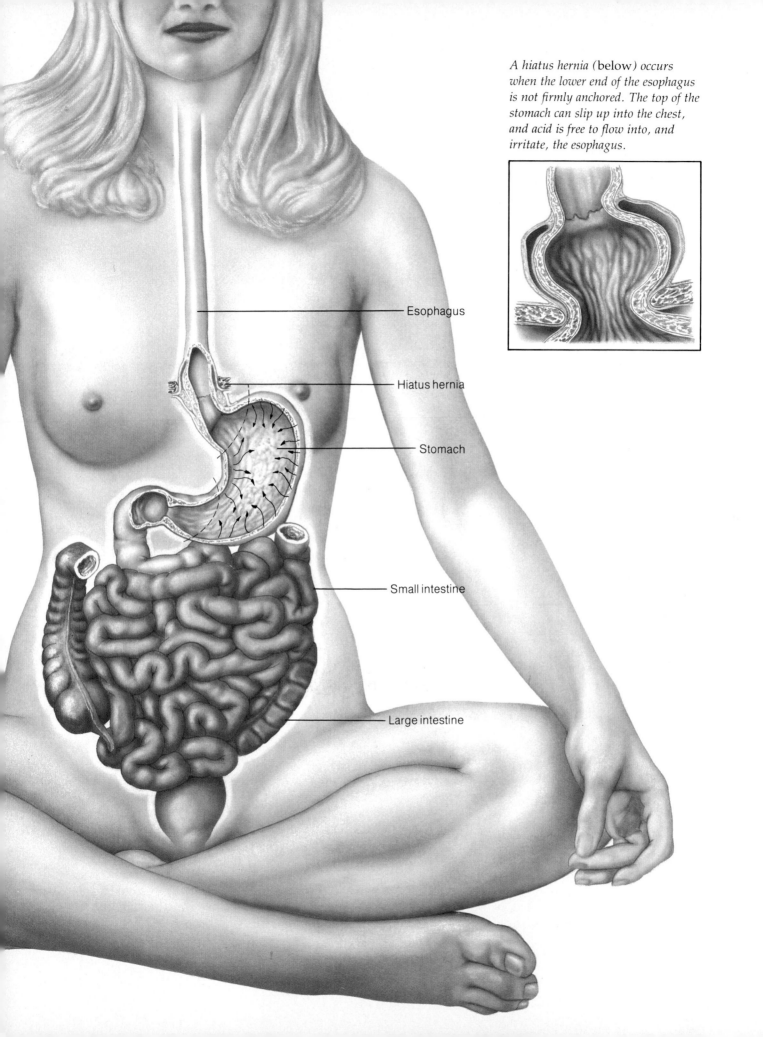

A hiatus hernia (below) occurs when the lower end of the esophagus is not firmly anchored. The top of the stomach can slip up into the chest, and acid is free to flow into, and irritate, the esophagus.

Esophagus

Hiatus hernia

Stomach

Small intestine

Large intestine

element, the disease being more frequent in those with certain blood group types.

Occasionally the production of the acid-stimulating hormone gastrin, runs out of control as a result of a gastrin-forming tumor, and acid production exceeds the preventive capabilities of even those with normal mucosal resistance to ulceration. In this syndrome, there are usually multiple and recurrent ulcers in the duodenum and even farther down the gut. Drs. Robert Zollinger and Edwin Ellison from the Midwest, while studying this disease, discovered that the production by a tumor in the pancreas of a gastrinlike substance over-stimulates the parietal cells of the stomach to produce a pint of highly acid fluid per hour. The treatment is complicated by the fact that pancreatic tumors are often multiple, and their position may be difficult to pinpoint for removal. Surgical removal of a part or all of the stomach may be necessary, though modern acid-suppressing drugs offer sufferers new hope.

Gastric ulcers are often associated with evidence of widespread inflammation of the stomach lining, causing breakdown of the mucosal resistance. Such gastritis may occur through drinking too much alcohol or through use of some drugs such as aspirin. Once formed, the ulcer will perpetuate itself and become chronic unless treated.

If the ulcer penetrates through all the layers of the stomach or the first part of the duodenum, perforation may occur, which was fatal for the ulcer sufferer in the days before safe abdominal surgery. The symptoms of perforation are dramatic. After some weeks of ulcer pain, the victim suddenly experiences severe upper abdominal pain. The person then rapidly becomes pale and shocked as the acid contents leak out into the cavity of the abdomen, the peritoneum. This acid peritonitis will rapidly be fatal unless surgery is performed quickly. Usually the surgeon simply sews over the hole, but, in some cases, he may choose to perform an additional operation to reduce the acid production of the stomach.

Before the recent introduction of fiberoptic gastroscopes, diagnosis of peptice ulcers depended on X-ray examinations of the upper gut. The more refined X-ray examinations now performed use barium salts to line the stomach and duodenum, where ulcers show as craters or deformation. However, as up to one-third of ulcers fails to show on X-ray examinations, gastroscopic diagnosis is increasingly used.

Surgery to curtail the acid production of the stomach or to remove peptic ulcers is now performed less often. This is due to the introduction, since the mid 1970s, of more effective chemical means of speeding up the healing of ulcers.

The most familiar medical treatments of ulcers are antacid pills or medicines. These are all alkali mixtures, which neutralize the stomach acid for a while. Again, over the last ten years more long-lasting reduction of the acid production of the stomach has become possible with the introduction of a series of drugs, including Tagamet and Ranitidine, which specifically stop acid secretion by the parietal cells.

Physicians used to be specially wary of gastric ulcers, since some of these turn out to be malignant, developing in time to form spreading cancers. Nowadays repeated gastroscopic examinations enable the physician to test for full healing of ulcers and to see that no cancerous change has occurred. Cancer of the stomach causes an estimated 15,000 deaths per year in the U.S.A. One quarter of the victims of this cancer — only one in ten survive for more than five years — start with symptoms identical to those of ordinary peptic ulcers. In some countries, notably Japan, stomach cancer is one of the commonest malignant growths. The cause is unknown, but the high proportion of smoked food in the diet has been implicated. Treatment is surgical and only succeeds where the tumor is at an early stage and has not spread to neighboring lymph glands or to other parts of the body.

The outlet of the stomach can become so scarred by chronic peptic ulceration that the pylorus is changed into a tight, unyielding obstruction to the passage of food to the duodenum. This pyloric stenosis causes foul vomiting and serious malnutrition as the stomach bloats with fermenting food which cannot be absorbed by the gut. Surgical unblocking of the pylorus is essential to prevent complications related to the loss of salts in the vomiting, as well as the nutritional losses. Distinct from this variety of pyloric stenosis, the outlet of the stomach may be overmuscular from birth, a con-

*Theodore Billroth published a
book on clinical surgery in 1881
in which were illustrated his
different techniques for the
surgical removal of cancers of the
duodenum.*

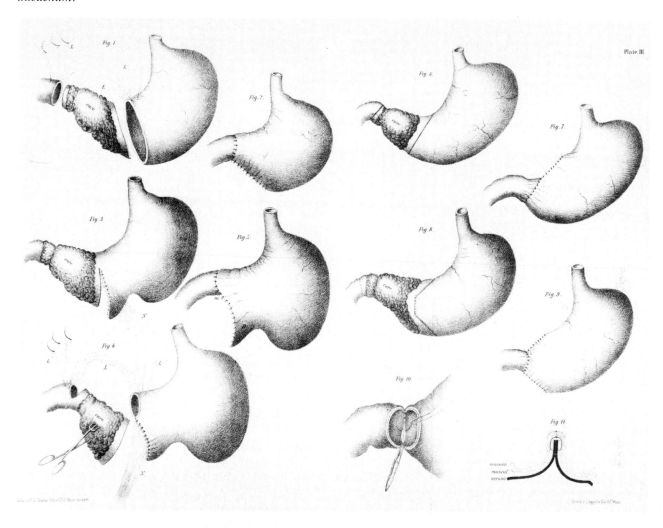

dition which usually shows itself in the first six weeks of life with profuse projectile vomiting. The problem is often seen in firstborn male children and is frequently genetically acquired. Again surgery is required, but usually of a simple kind, in which the muscle fibers are slit and resewn at right angles to reopen the pyloric passage to the duodenum.

The vomiting of blood is an alarming symptom to both the sufferer and the physician. Bleeding from the upper gut usually occurs as a result of peptic ulceration penetrating the wall of a blood vessel in the lining of the stomach or duodenum. More serious than this is the bleeding that occurs from veins in the esophagus near its junction with the stomach when these are made varicose or swollen by disease in the liver. Bleeding from these esophageal varices is often torrential and is frequently the cause of death in alcoholic cirrhosis. Urgent gastroscopy is essential to show the cause of the bleeding, and sometimes to treat it by coating the ulcer with a cementlike substance or with laser beams. When the site of the broken blood vessel is not clear, even after gastroscopy, arteriography may be necessary. Here the major arteries to the stomach are cannulated, and dye opaque to X rays is injected, so showing up the leak into the upper digestive track. When the bleeding point is in a place inaccessible to a surgeon, the damaged artery can be plugged through the same cannula — tube — from which the X-ray dye was injected, using special quick-setting resins to cement the blood vessels shut.

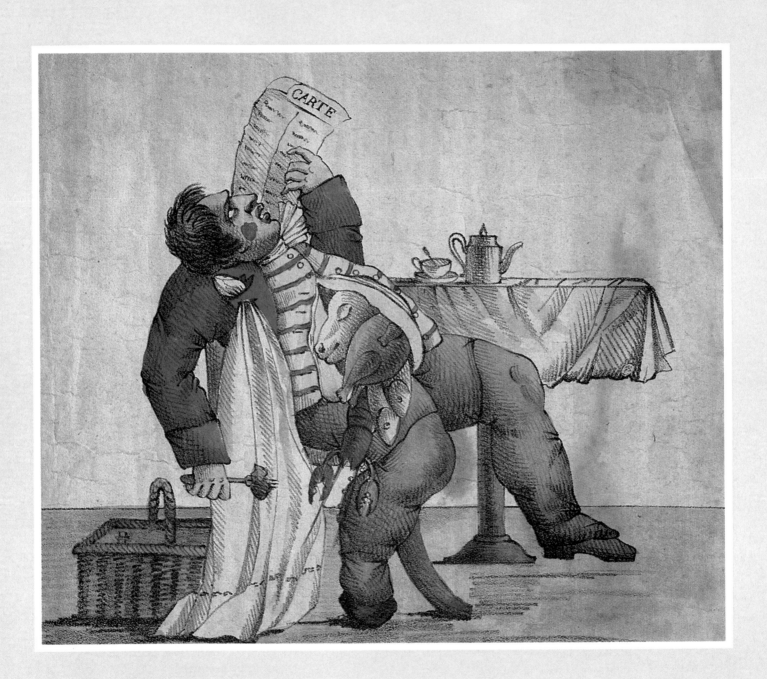

Chapter 4

Breaching the Barrier

Throughout the course of medical history, there have been periods when the resources of medical science have been concentrated on a particular organ òr tissue with particularly fruitful results. The small intestine and the associated organs of digestion, such as the stomach, the pancreas and the biliary system, seem to have enjoyed special attention during the last decades of the nineteenth century and the first few years of the twentieth century. An extraordinarily powerful group of minds concentrated on the question of the function of the digestive secretions and the mechanisms controlling them.

Two of the greatest experimental medical scientists of the nineteenth century, the Frenchman Claude Bernard and the Russian Ivan Pavlov, began their experimental careers as a result of an interest in the control of the digestive juices. Later, in 1902, Bayliss and Starling, the possessors of minds of a power to rival those of Bernard and Pavlov, made a crucial step forward. They demonstrated that the secretion of digestive juices was partly controlled by a substance carried in the bloodstream. The interesting feature of this substance was that it caused its action at a site remote from its actual point of secretion, or manufacture. They called the substance they found "secretin," and they demonstrated that it controlled the production of the pancreatic juices. Although this discovery was of vital importance, the idea behind it was even more important. Bayliss and Starling called their chemical messenger a "hormone." Subsequently, research has uncovered many other hormones, responsible for stimulating actions throughout the body, but the findings of Bayliss and Starling pioneered the fascinating field of hormone science, today called endocrinology.

The discovery of secretin by Bayliss and Starling was the first in a long line of similar discoveries. Physiologists now know of more than fifteen different substances secreted by one part of the

This French satirical sketch, L'Indigestion Anglaise, *may seem extreme today, but in the early nineteenth century huge meals were the order of the day for the rich, and indigestion was a frequent problem.*

gastrointestinal system which have an effect upon another. All these different compounds are made up of strings of amino acids — polypeptides. The major technique for identifying these biochemical substances is called radioimmunoassay (RIA). It is a sensitive technique, but it cannot be relied upon with certainty to differentiate between closely related compounds so that it is impossible to know exactly how many gastrointestinal hormones exist. Also many of the different substances have closely related effects, and a single substance may have an effect upon one part of the system in a small concentration and upon other parts of the system in much higher concentrations.

As well as being influenced by the gastrointestinal hormones, the function of the digestive system is also affected by the nervous system. This is particularly important in modulating the activity of the stomach, and less important in controlling the secretions of the pancreas and the gallbladder. The control of the substances that are liberated from the wall of the duodenum itself is probably exerted exclusively by the nervous system, but the system works in such a way that very local reflex arcs are involved, using only the nerves within the wall of the duodenum itself. The brain and the spinal cord are not involved at all.

Gastrin is one of the most important of the gastrointestinal hormones. It is produced in the wall of the duodenum and in the wall of the stomach in its lowest parts. In fact it is not a single substance but a group of closely related substances, all of which seem to have similar effects. The main effect of gastrin is to stimulate the production of acid by the stomach wall and to increase the blood flow through the stomach.

Cholecystokinin and secretin are also produced in the duodenal wall. Not only do they trigger secretion by the pancreas and the gallbladder, but they also inhibit the emptying of the stomach. Pancreozymin and bradykinin also come from the duodenal wall. Pancreozymin stimulates an enzyme rich pancreatic secretion, while bradykinin is involved in increasing blood flow.

A picture emerges of an extremely elegant and complex system based on the secretion of a group

Ernest Henry Starling

Discovery of the Chemical Messenger

Ernest Henry Starling, known as the "clinician's physiologist," is remembered as one of the greats of this discipline. His name is familiar to all who have undergone medical training and have learned Starling's Law, which concerns heart muscle contraction, but he researched widely, particularly in the field of gastrointestinal hormones.

Starling was born on April 17, 1866 in North London and entered Guy's Hospital to study medicine in 1882. He was a distinguished student and went on to start work in the Physiology Department at Guy's in 1889. The department was poorly equipped, and Starling was forced to carry out some of his research at University College, London, where he met (Sir) William Maddock Bayliss, with whom he collaborated extensively.

Starling was a man of considerable energy, alert and questioning, with the ability to make rapid decisions. In 1899 he became the Jodrell Professor of Physiology at University College, and in the same year was elected a Fellow of the Royal Society. In 1912, he published the classic physiological textbook, *Principles of Human Physiology.* His first important researches were concerned with lymph flow and were performed in Breslau. His work on the formation and flow of tissue fluids provided new

insights which have governed thinking in the field to this day.

In the early years of the twentieth century, Starling worked on pancreatic secretion. This work was done with Bayliss and, in 1902, they discovered the hormone, secretin, which stimulates pancreatic secretions. Now that we are familiar with the concept of hormones, it is difficult to understand how new and exciting was this idea of an internal secretion which had a specific action on a separate organ, the pancreas. Epinephrine had been discovered eight years earlier, but its effects were widespread. Essentially Starling's greatest contribution to the world of medicine was to define a hormone —
a substance produced by one tissue which had an effect on another tissue.

Next, he turned his attention to the heart, the subject of some

of his earlier researches. He propounded his law of the heart, stating that "the energy of contraction is a function of the length of the muscle fibers." That is, the more the heart is filled with blood, the more forcefully will it contract.

During World War I, Starling worked on gas warfare, and in 1917 was sent to Italy to persuade the authorities there to use an efficient respirator for their troops. Respirators were supplied to the Italian army by the British and saved many lives during Austrian gas attacks.

Starling received many accolades for his work: the Royal Society's medal in 1913 and honorary doctorates from the Universities of Sheffield, Cambridge, Breslau, Strasbourg and Heidelberg, and from Trinity College, Dublin.

Before his death in 1927, Starling summed up his life and work in his book *A Century of Physiology* by saying: "When I compare our present knowledge of the workings of the body with the ignorance and despairing impotence of my student days, I feel that I have had the good fortune to see the sun rise on a darkened world, and that the life of my contemporaries has coincided not with a renaissance, but with a new birth of man's power over his environment and his destiny unparalleled in the whole history of mankind."

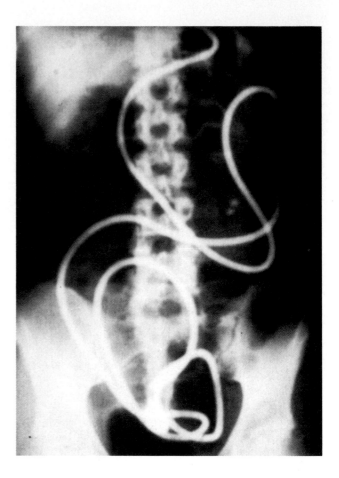

of substances which are often very similar in their chemical structure.

The end result is a subtle control mechanism which maintains a smooth flow of food through the system, backed by appropriate secretion of enzymes, protective mucus and digestive juices, and an increase of blood flow. In the past, it was thought that there was a distinction between the action of the various hormones and the effects of the nervous system. However, physiologists are now beginning to get a few tantalizing glimpses of the completeness of the system, and it is becoming clear that some of the substances that appear to act as gastrointestinal hormones also act as nerve transmitters deep within the brain.

The pioneering investigators, Bayliss and Starling, were closely involved in discovering the chemical control mechanisms of the body. Not only was this chemical approach to the body's activities essential to the foundation of endocrinology, but it also did a great deal to direct physiology into the area of metabolic medicine. Metabolic medicine is, in fact, the investigation of the body's activity from a chemical point of view. The chemical and metabolic approaches to the study of digestion have shown just how complicated the system is of breaking down our food into basic chemical building blocks, which can be absorbed easily and then used by the body to maintain life.

Enzymes — the Workhorses of Digestion

While the acid secreted in the stomach does a great deal to break down food, most digestion actually takes place in the small intestine. Liquefied food, or chyme, flows into the duodenum (the first part of the small intestine) where it is mixed with large amounts of enzyme containing alkaline fluid from the pancreas. This fluid acts to neutralize the effects of the stomach's acid. The enzymes in chyme, which are the "sharp end" of the digestive process, are produced by both the pancreas and by the wall of the intestine. Essentially, enzymes bring about the reduction of various food constituents, such as starches and fats, into "chemical pieces" small enough to be absorbed by the intestine. But what exactly are enzymes?

Enzymes are large protein molecules, and they act as catalysts by triggering and speeding the chemical reactions involved in digestion. Many of these would happen too slowly to be of use without enzymes, whose role in speeding up the reactions to a fast enough rate is crucial. Beaumont showed early in the history of enzyme research that there were substances in the gastric juices that were capable of digesting meat. Later, in 1835, the German chemist Theodor Schwann demonstrated the existence of the enzyme pepsin. It was this that initiated the proper study of enzymes, for they are found freely within the stomach and intestine and so are relatively easy for scientists to obtain. It is now realized that nearly all the complex chemical processes of the body are both initiated and controlled by the activity of enzymes.

The first enzyme food meets is in the mouth. This enzyme, salivary amylase, is produced in small quantities by the salivary glands. It acts on some of the starch present in the food and starts to break it down into smaller molecules. Although this enzyme ceases to act in the acid conditions of the stomach, its effect continues until the acid juices penetrate the heart of the ball of food that was formed in the mouth.

The next enzyme encountered by food is pepsin. It is found in the stomach and is partly responsible

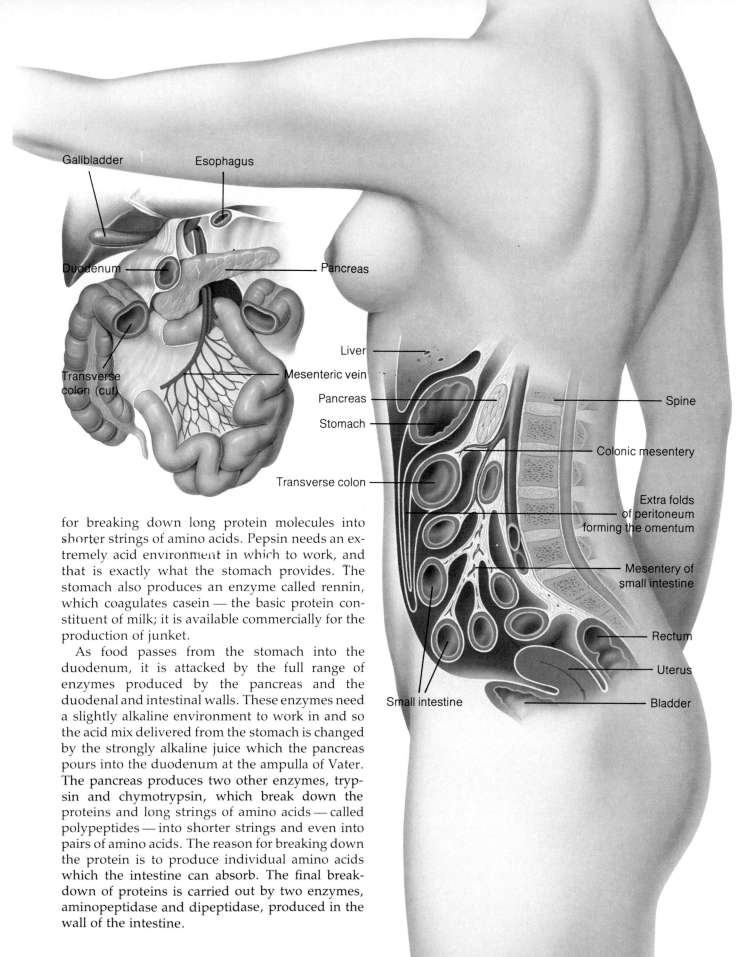

Gallbladder

Esophagus

Duodenum

Pancreas

Transverse colon (cut)

Mesenteric vein

Liver

Pancreas

Spine

Stomach

Colonic mesentery

Transverse colon

Extra folds of peritoneum forming the omentum

Mesentery of small intestine

Rectum

Uterus

Small intestine

Bladder

for breaking down long protein molecules into shorter strings of amino acids. Pepsin needs an extremely acid environment in which to work, and that is exactly what the stomach provides. The stomach also produces an enzyme called rennin, which coagulates casein — the basic protein constituent of milk; it is available commercially for the production of junket.

As food passes from the stomach into the duodenum, it is attacked by the full range of enzymes produced by the pancreas and the duodenal and intestinal walls. These enzymes need a slightly alkaline environment to work in and so the acid mix delivered from the stomach is changed by the strongly alkaline juice which the pancreas pours into the duodenum at the ampulla of Vater. The pancreas produces two other enzymes, trypsin and chymotrypsin, which break down the proteins and long strings of amino acids — called polypeptides — into shorter strings and even into pairs of amino acids. The reason for breaking down the protein is to produce individual amino acids which the intestine can absorb. The final breakdown of proteins is carried out by two enzymes, aminopeptidase and dipeptidase, produced in the wall of the intestine.

However, there is a problem concerning the manufacture of these powerful protein-splitting enzymes since they are produced in the pancreas, which, like the rest of the body, has a high proportion of protein in its structure. If these enzymes were excreted within the pancreas in an active form, they might very well digest the pancreas itself, producing a hole in either this gland or the duct joining the pancreas to the intestine. Conveniently, the protein-splitting enzymes are secreted in an inactive form, which is converted into its active mode as soon as the pancreatic juice reaches the intestine. As an example, the enzymes trypsin and chymotrypsin are secreted in the pancreas in the form of the inactive compounds trypsinogen and chymotrypsinogen. It takes yet another enzyme — an activating enzyme called enterokinase which is made by cells in the duodenal wall — to enable trypsinogen and chymotrypsinogen to be converted into their active forms when they enter the duodenum.

A subtle additional step in the digestion of fat by enzymes is needed before the process can take place effectively. Because fat and water do not mix, fat tends to sit in the intestine in globules which the fat-splitting enzyme, lipase (produced by the pancreas), finds hard to penetrate. These globules must, therefore, be broken down into much smaller droplets — a process which chemists refer to as emulsification. This is effected by bile, which is released from the gallbladder as food enters the duodenum from the stomach; it is an especially good emulsifying agent whose basic function is to split fat globules into droplets small enough for the lipase to act upon. There is also a cholesterol-splitting enzyme present in pancreatic juice that splits up any free cholesterol in the chyme.

Finally, the breakdown of large carbohydrate complexes, such as starch, although partly brought about by the amylase in saliva, is mostly performed by a similar enzyme, released from the pancreas, called pancreatic amylase. As a result of the actions of these two enzymes, most food starch is broken down into simple sugars such as the disaccharides and monosaccharides, especially one called maltose. Further breakdown of the disaccharides into the smaller monosaccharides takes place in the wall of the small intestine as a result of the action of a

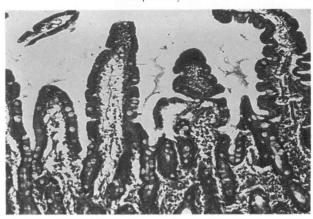

number of specific enzymes such as maltase, sucrase and lactase. The shape, size, function and workings of the small intestine are all matters of interest to scientists and laymen.

Absorption — a Tennis Court Packed into a Tube

The small intestine — a tube about twenty-five feet long and about one inch in internal diameter — is something of a miracle of biological packaging in the way it is coiled up and fitted into the abdominal cavity. From the anatomical point of view, this long tube is divided into three named parts. The first twelve inches is called the duodenum — a word which simply means twelve inches in Latin. This is followed by the jejunum, which is about ten feet long; the final fifteen feet or so is made up of the thinner walled ileum.

All this tubing sits in the abdomen in folds. But it does not float freely about within the abdominal cavity since there would then be every chance that these folds would become twisted together and cause a blockage. Instead, the intestine is anchored to the rear wall of the abdomen by a structure called the mesentery. This thin, membranous sheet not only holds the intestine in place but also carries blood vessels that are essential for both the supply of oxygen to the intestine and the transport of absorbed nutrients away from it to the liver and the rest of the body. The appearance of the mesentery has been likened to an open fan, rooted in the back wall of the abdomen, with its long, free edge supporting a great length of small intestine.

The remarkable nature of the intestine's structure soon becomes clear when the internal surface of

The internal surface of the intestine, is designed for maximum absorption. The mucosa itself is gathered up into epithelial folds. The surface of the mucosa, when viewed under a microscope, shows numerous fingerlike projections, called villi. The absorptive cells — the enterocytes are ranged on the surface of the villi. Each individual cell has its own raised projections — the microvilli. The combination of epithelial folds, villi and microvilli increases the absorptive surfaces of the intestine to an area 200 times larger than that of the entire skin that covers the human body through which nutrients are taken in.

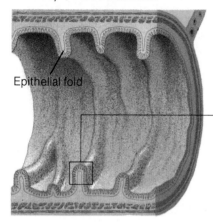

Epithelial fold

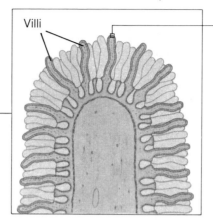

Villi

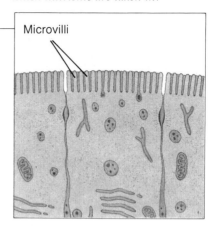

Microvilli

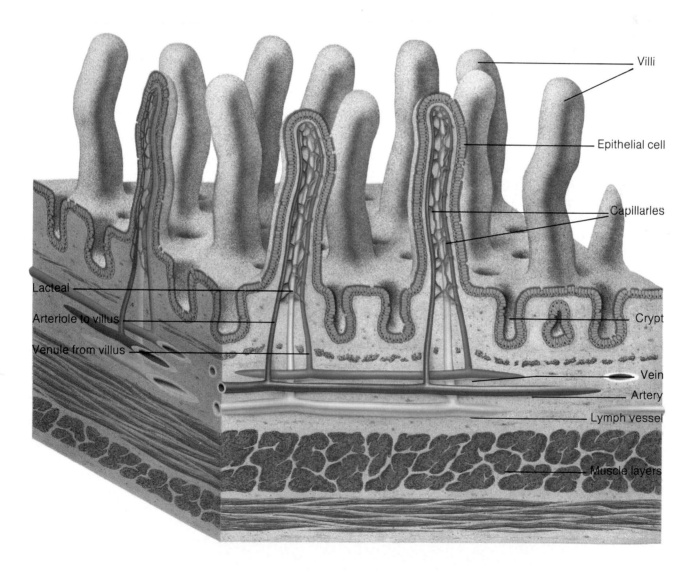

Villi

Epithelial cell

Capillaries

Lacteal

Arteriole to villus

Venule from villus

Crypt

Vein

Artery

Lymph vessel

Muscle layers

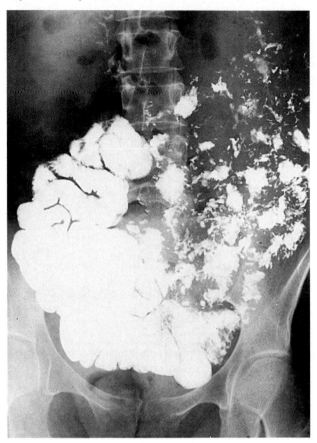

If the small intestine were a simple, featureless tube, then it would have a surface area of over four square yards. However, the combination of the folds, the villi and the microvilli greatly increases the intestine's area for absorption to about three hundred square yards — larger than the size of a tennis court. This huge area is not only important in the facilitation of absorption but also to the work of digestion, since many of the enzymes actually work at the level of the microvilli.

Although we think of our alimentary canal as being very much "inside" ourselves, it is, of course, in direct continuity with the external environment. In fact, the lining of the intestine is one of the important areas of the body where direct interface with the outside environment occurs. Nutrients or, indeed, potential invading organisms do not really become a part of the body until they have passed through the layer of cells that forms the inner lining, or epithelium, of the intestine.

These highly specialized cells are created and shed over a relatively short space of time. They start their life in the many "crypts" which surround each villus. They migrate up the wall of the crypt and then onto the villus itself, rather as though they were on an escalator, before finally being shed from the tip of the villus. While they are still within the crypt, the cells are immature and not capable of fulfilling either their digestive or their absorptive function. They reach maturity as they start to ascend the villus. So fast is the rate of turnover that they have an active life of only three or four days before they are shed into the lumen of the intestine.

Once food has been broken down into sufficiently small chemical constituents to be passed through the intestinal epithelium, it is picked up by capillaries, small blood vessels which eventually drain into the portal vein. Each of the villi has its own network of capillaries so there is no shortage of channels to supply and then drain the nutrient rich blood away from the intestine. All the nutrients absorbed from the intestine are carried in the blood through the portal vein to the liver, where further chemical rearrangement takes place so that the body's overall nutritional needs can be met. There are a number of different ways in which substances can be carried across the intestinal wall so that they enter the bloodstream. Water is one of the most

this lengthy tube is considered. It is on this surface, known as the intestinal mucosa, that nutrient absorption takes place. To the naked eye, it appears that the mucosa is gathered up into folds, and these folds increase the surface area available for absorption. However, if the surface is looked at under a magnifying lens it looks velvety; and a microscope will show that this velvety appearance results from numerous fingerlike projections, called villi, which protrude into the lumen, or central section of the intestinal tube. These villi increase the internal surface area even further.

But the story does not end there. Once the level of magnification is turned high enough to scrutinize a single cell, it can be seen that it too has a surface raised up into fingerlike projections, known as microvilli. At every step of this journey of increasing magnification, the intestine increases its surface area so that the maximum possible area is available for the absorption of food.

important substances that the intestine absorbs, and this passes across by a process known as passive diffusion. The contents of the intestinal lumen are relatively dilute, that is to say there is a large amount of water in them. In contrast, the contents of the intestinal cells contain large amounts of chemicals of various types, dissolved in a small amount of water. This discrepancy in the strength of the two solutions on the different sides of the cell membrane brings the force of osmosis into action and passively "sucks" water from the intestinal lumen into the intestinal cells.

Other substances, such as the essential mineral sodium, have to be absorbed by an "active transport mechanism." This means that there is an energy-using process going on whereby special chemical "pumps" in the cell wall carry sodium into the cell. Every sodium ion that is absorbed needs the expenditure of energy by the intestinal cell to transport it from the intestinal lumen into the cell.

It is as though the cell is pushing the sodium uphill. There is a higher concentration of sodium within the cell than outside, and natural forces, similar to the forces of osmosis, will tend to move the sodium the other way. It is only by expending energy that the intestinal cell overcomes these forces. Amino acids and sugars are other substances absorbed by the process of active transport.

Fat Absorption

Absorption into the blood stream works well for the uptake of sugars and amino acids — the breakdown products of starch and proteins. There is some difficulty, however, in organizing the handling of fats because they do not dissolve in water. The manner in which a fat (lipid) is absorbed depends upon the length of the carbon chain that makes up its constituent fatty acids. Those with short chains can be absorbed by the intestinal cells. But the majority of lipids have to be absorbed by one of the body's

cell directly into the portal circulation in the same manner as the carbohydrates and the proteins, but are taken up by the lymphatic system instead. The lymphatics are a system of tiny vessels, similar to the blood vessels, that are found throughout the body. The function of the lymphatic system is closely tied in with the body's immune defense system, since the lymphatics pass through lymph-nodes, which house the cells that control and trigger the body's response to invading organisms. And the role of the lymphatics in handling fats is secondary to its role in the defense of the body.

The tiny lymph channels that supply each individual villus are called lacteals; they eventually link up to form a major lymphatic channel in the chest called the thoracic duct. This duct empties into the circulation at the point where the main veins of the left side of the neck join together before entering the heart.

Malabsorption — the System Breaks Down

The small intestine only rarely gives rise to any sort of trouble; it is an organ which seems to carry on fulfilling its function in an uncomplaining fashion almost indefinitely in most people. However, there are exceptions.

If trouble occurs, then the usual manner in which it comes to light is as a disturbance in the way in which the intestine breaks down and absorbs food. Any such disturbance is known as malabsorption, which can show itself in a variety of ways, but usually a disturbance in the way in which the intestine handles fats reveals that something is amiss. This is hardly surprising, since the system for digesting and absorbing fats is much more complex than that involved in the absorption of proteins or carbohydrates.

Malabsorption of fat leads to the production of very bulky and offensive-smelling stools which tend to float in the toilet bowl and can be difficult to flush away; these strange stools result from their high content of malabsorbed fat. Another symptom is weight loss, as a result of a failure of nutrient supplies. Other likely problems are anemia, since the absorption of iron and folic acid (one of the B vitamins) falls off and causes disturbances in the blood level of calcium — discovered by a blood test. Calcium levels are upset because vitamin D, a

more complex mechanisms, involving bile salts from the gallbladder and the enzyme lipase from the pancreas. They are first broken down into their constituents — fatty acids — by the action of pancreatic lipase, and they then form a water soluble complex with bile salts secreted into the intestine by the gallbladder. This complex is called a micelle. The micelles are carried to the surface of the intestine, where they break down to liberate free fatty acids which are able to pass across the intestinal cell membrane and on into the cell. No energy is used in this process since the cell membrane is also fatty in nature. The absorbed fats merge with the cell wall and are then assimilated into the cell itself. The bile salts are left behind in the intestinal lumen to form more micelles, and, when there is no more fat to be absorbed, they pass on down the intestine to be reabsorbed in the final part of the ileum. They get back to the liver and the gallbladder by way of the bloodstream.

Once the fatty acids are within the intestinal cell, they are formed into triglycerides and "wrapped up" in a protein coating, which makes them soluble in water. These tiny packages of fat are known as chylomicrons. They do not pass from the intestinal

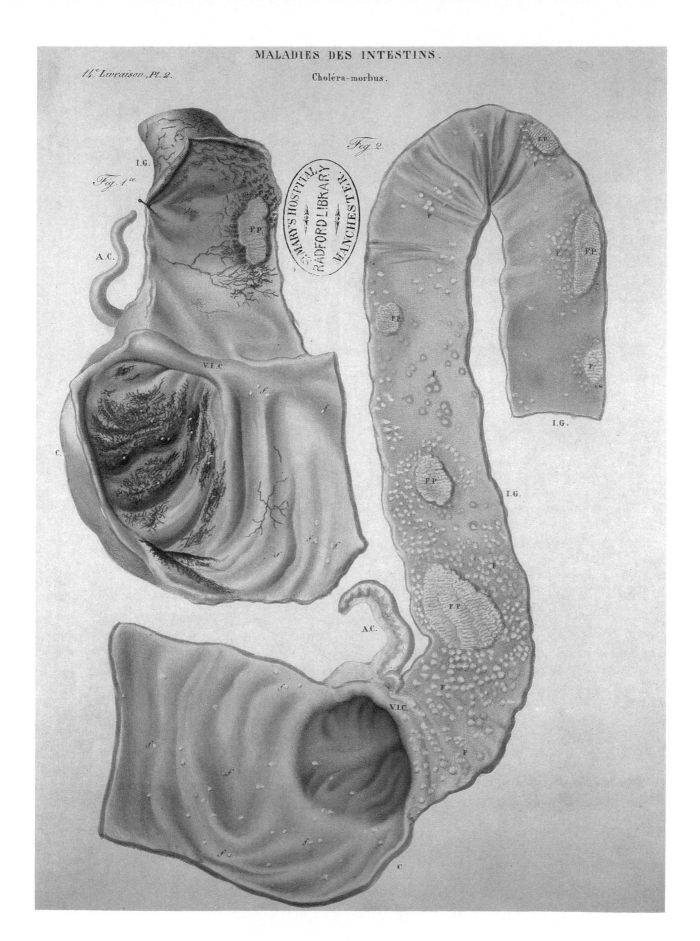

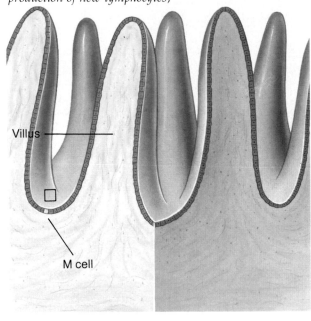

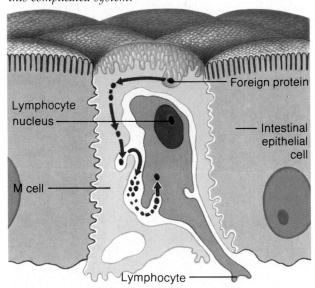

fat soluble vitamin, cannot be absorbed properly. Malabsorption can be caused by a large number of different complaints, but the most important are perhaps those which result from diseases which affect the intestinal epithelial cells themselves. Without doubt, the most important of these is a disease called celiac disease, or gluten enteropathy, which results in diarrhea, abdominal pain and weight loss.

Celiac Disease

Tracing celiac disease and its cause involved the diet in the war-torn Holland of the 1940s where bread was in very short supply. This lack of bread was partly the result of Hitler's policy of trying to starve the Dutch, since they had shown an uncommon degree of resistance to the forces of Nazi oppression. In any event, a physician named Dicke working in Holland at this time, noted that many of his patients with celiac disease seemed to get better rather than worse as a result of this relative degree of starvation. In Sweden, at the same time, there was much less of a shortage and, strangely, patients with celiac disease in Sweden showed no improvement in their condition during this period. Dicke subsequently showed that celiac disease is caused by the substance gluten, which is present in wheat and, indeed, in most cereals.

Celiac disease results from hypersensitivity of the small intestine to the gluten present in wheat. It is essentially an allergic reaction, with the intestinal cells falsely perceiving the gluten to be an invading organism, rather than a normal constituent of the diet. This sets up the pattern of continuing inflammation in the wall of the intestine.

One striking feature of celiac disease is only visible under the microscope. When the mucosa is observed, it can be seen that all the villi have been lost and the surface is flat and relatively featureless. This picture under the microscope is the hallmark of celiac disease, and it, therefore, becomes essential to have some way of obtaining samples of the mucosa for microscopic examination. In 1958, W.H. Crosby, an American physician, developed a most ingenious device for obtaining just such a sample. The device consists of a small capsule on the end of a tube which can be swallowed. It has a single orifice, which comes to lie against the surface of the intestinal mucosa. If a vacuum is then applied to the end of the tube, a segment of mucosa is sucked into the capsule, and the vacuum operates a small, swivel-based knife which neatly clips off the mucosal segment. With its cargo safely on board, the capsule can be pulled back up through the stomach and out of the mouth.

This technique, called intestinal biopsy, is not

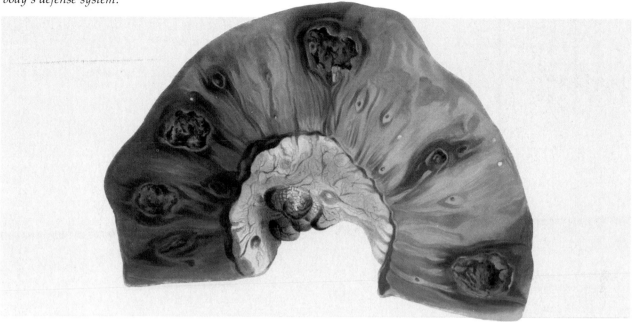

only necessary to make the diagnosis of celiac disease, it also provides an excellent way of following the progress of the disease. Once celiac disease has been diagnosed, patients should adhere to a diet containing no gluten, which means a diet more or less without food of cereal origin. Repeated biopsy gives both physician and patient some idea of how successfully the disease is being controlled.

Usually celiac disease is diagnosed in the first year of life, following weaning onto gluten-containing cereals. However, today there seem to be almost as many cases of celiac disease diagnosed later on in life, and it is not uncommon for the condition to be discovered when a person is forty or fifty years old. A retrospective analysis often suggests that these older patients may have been suffering from the disease for many years before diagnosis. There is now a small amount of evidence to suggest that early weaning onto cereal foods is not a very good idea and that avoiding gluten-containing food for the first four or five months of life may reduce the risk of celiac disease developing in infancy.

Bacteria and the Bowel

Everything in the world outside the body contains bacteria, and this includes the food we eat. As food descends into the acid bath of the stomach, many of the bacteria are destroyed. The food that passes on into the intestine, therefore, contains only a low count of bacteria, and this remains low until food reaches the colon, where there is a large concentration of bacteria present.

Not all the bacteria in the colon are harmful. For example, a number of the vitamins essential for health are synthesized by bacteria in the large bowel and then absorbed for use within the body. In fact, most of our daily needs of the two B vitamins pantothenic acid and biotin probably come from this source, as does most of the vitamin K needed for normal blood-clotting.

The Intestine's Defence System

Knowledge of the immune defense system operating within the body has advanced as fast as any other area of medical knowledge in the last few decades. The picture that emerges shows a closely interwoven system, involving all the different types of white blood cell, the antibodies that circulate in the bloodstream, and the lymph nodes and other patches of lymph tissue within the body — all fighting the defensive fight.

In the year 1677, however, the concept of a unified defense system against external invaders would have been totally alien to the Swiss doctor Johan Peyer. When he described the patches of lymphatic tissue in the small intestine which now

mounting the defense against invasion by bacteria and other organisms. They work by triggering the production of large numbers of new lymphocytes, capable of making antibodies, which will attach themselves to the invading organism.

These new lymphocytes are made within the Peyer's patches themselves and in the lymph nodes that lie in the base of the mesentery; but to get to where they are needed, they have to travel a circuitous route. Although they will eventually return to the intestinal mucosa, from Peyer's patches and the lymph nodes they pass into the thoracic duct and, from there, into the circulation, where the duct joins the venous system. Once in the circulation, they are carried back to the area where antibody production is needed. One of the body's continuing mysteries is the mechanism that controls the lymphocytes' return to the relevant part of the intestinal mucosa.

The antibodies manufactured by these newly made lymphocytes are large protein molecules, designed to latch onto foreign protein molecules like a key fitting into a lock. Within the body there are four main classes of antibody, or immunoglobulin, known by the initials IgG, IgA, IgE and IgM. The antibody that acts on the surface of the intestine, like a sort of antiseptic paint, is IgA.

Compared with the rest of the body, there are particular problems relating to antibody action in the intestine, as a result of the circumstances found there. Thanks to its role in digestion, the intestine is full of protein-splitting enzymes, so the IgA molecules need to be protected from the effects of these enzymes. This is done by the production of a special type of "handle" — known as a "J" piece — to which two IgA molecules bind. This combined molecule is called "secretory IgA," and it is resistant to the effects of digestive enzymes.

The body's protective apparatus works by being able to recognize certain types of protein as "foreign." The remarkable thing about the system is the fact that there is a continuous load of foreign protein, in the form of food, being emptied into the intestine all the time. The system of M cells and lymphocytes in the intestinal wall seems to make a good job of differentiating between the foreign, but acceptable, food protein and the unacceptable protein in the walls of invading organisms.

bear his name, Peyer thought they were involved in some sort of secretion. However, their association with infection became clear during the next century because post mortem studies showed them to be the major site of infection in typhoid fever.

Although Peyer did not know it, his patches are an important part of the way in which the small intestine protects itself from invasion by bacteria. The patches contain large numbers of lymphocytes — the key cells in the immune system; they also contain a newly discovered special sort of cell called an "M," or "microfold," cell. These M cells seem to attract bacteria and other toxic substances to themselves and then funnel them down to make contact with a lymphocyte. This is a cunning system, for lymphocytes are responsible for

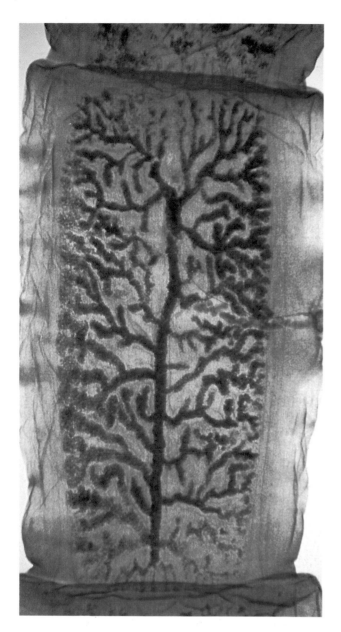

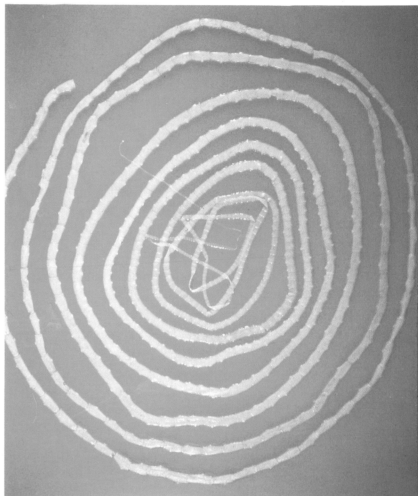

The Taenia saginatus *is found in the intestine of the cow. The adult tapeworm produces eggs which are excreted in the feces, and if this is used as manure, vegetables can be contaminated.*

Occasionally, of course, the system slips up. Celiac disease, for instance, results from a misinterpretation of gluten as a potentially harmful substance, and there are other food allergies which come about in the same way.

Typhoid Fever

The habit cooks had of moving from job to job was legendary in the early part of this century and Mary Mallon was no exception. She moved from household to household in the richer parts of New York, holding eight jobs in seven years. Typhoid fever struck seven of those eight households.

Mary Mallon, or "Typhoid Mary" as she was known, is the best publicized example of a carrier of typhoid fever. She herself had probably suffered no more than minor attacks of diarrhea when she acquired the disease, and the hardy typhoid bacillus came to rest in her gallbladder where it lived quietly and happily, excreting large amounts of typhoid bacilli into her intestine and so into her feces. Typhoid Mary's crop of bacilli were greatly aided by her choice of profession, since any catering worker has the best possible opportunity to pass on the infection. The bacilli invade the Peyer's patches in the intestine and then go on to cause a general infection of the blood stream. At this stage of medical knowledge, in the early 1900s, treatment was poor; the patient would either die, as the blood became infected, or recover spontaneously.

When typhoid mysteriously broke out in Oyster Bay, Long Island, Dr George Soper from the Public

87

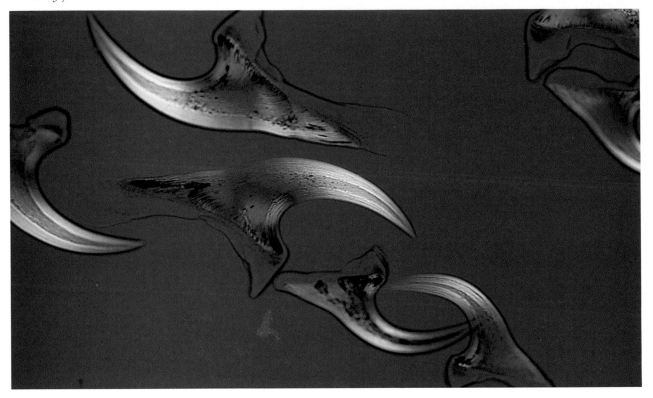

Health Department came to investigate. He heard about the cook (Mary), whose ice cream had been so delicious and who had left as soon as the typhoid appeared. Eventually he traced Mary, who by this time had moved through two further jobs and caused three more cases of typhoid. He found her in the kitchen of her latest employer and tried to suggest tactfully that she might have something to do with the outbreaks of typhoid that seemed to cling to her apron strings. She attacked him with a large meat cleaver.

Mary was next forcibly abducted to the Riverside Hospital on North Brother Island. Dr Soper showed that her feces were pervaded by typhoid bacilli and explained that she could be cured of her affliction — which caused far more harm to others than it did to her — by the removal of her gallbladder. Mary refused, not unreasonably, since cholecystectomy, as the operation is known, was at that time hazardous. She would not abandon her cooking either, so there was no option but to keep her prisoner at the hospital. In 1910, she promised she would work as a laundry maid, so she was released.

Despite her promise to report every three months, she disappeared.

In fact, she had become Mrs Brown — a cook. She moved from job to job, taking her ice cream recipe and typhoid fever with her. It was a common disease then, and not every outbreak was subjected to particular scrutiny. However, when twenty-five cases of typhoid occurred at the Sloane Hospital for Women in 1915, Mary was tracked down and detained by the Health Department. This time she was not allowed back out to cook her killer creations in the kitchen, and she spent the rest of her life on North Brother Island.

Thanks to Typhoid Mary and others like her, typhoid carriers are now sought with great diligence if there is an outbreak of the disease. However, the main preventive measure is a clean water supply, and typhoid ceases to be a major health hazard once this has been established, although occasional cases can still occur. Unfortunately, the disease is still endemic in many parts of the world, but there is also an effective vaccine which will provide protection for travelers.

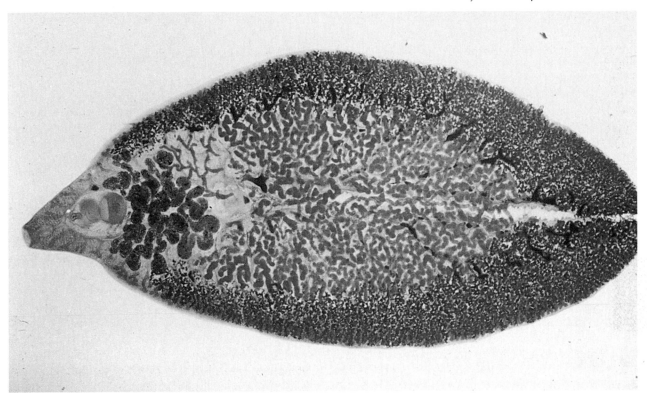

The liver fluke, Fasciola hepatica, *is usually found in sheep. However, humans can be infected through eating wild watercress. When this happens, the liver becomes enlarged and fever develops.*

Typhoid is caused by *Salmonella typhi* and is one of the family of salmonella organisms, all of which can cause infection of the intestine. However, typhoid is the only type of salmonellae to pass through the intestine and cause a bloodborne infection. The other members of the family cause a brisk diarrheal illness that usually causes little trouble in adults but can be very serious, or even fatal, in babies, young children and the elderly.

This is because of the role of the small intestine in the water balance of the body.

The Small Intestine as a Barrier

The mucosa of the small intestine has a hard task to perform. While it needs to facilitate the absorption of nutrients in every way, it still has to provide an effective barrier to the infections that occur in the outside world. This action is especially important in infants. Indeed, the major factor affecting the difference in infant mortality rates between developed and undeveloped countries is the incidence of infantile gastroenteritis.

One of the major tasks of the small and the large intestine, working together as a unit, is to prevent the loss of large amounts of body water, and the loss of vital minerals that body water contains. This ability can cease when gastroenteritis occurs. In an attack of diarrhea the excessive fluid loss is all too obvious. While an adult can temporarily afford to lose a fairly high percentage of his or her body water before being in danger of shock and collapse of the blood circulation, the same does not apply in an infant. Because the total circulating volume of body water in an infant is so small, a loss of even ten percent can prove fatal.

Giardia, a Protozoal Infection

The protozoa are another type of invading organism; they are single celled animals and larger than bacteria. The best known protozoal disease is malaria. The protozoan *Giardia lamblia* is one that commonly infects the upper part of the small intestine causing diarrhea. Like typhoid, this organism is almost always spread through infected water supplies and, again, in many parts of the world it is endemic, especially in fairly hot climates. It is not an exclusively tropical and subtropical disease though, and Leningrad, in the U.S.S.R., is well known as a giardia blackspot. Fortunately the disease responds well to treatment with a drug called metronidazole.

Organisms of the size of the typhoid bacillus and giardia can be controlled by the activities of the immune system. However, the intestinal parasites such as tapeworms, hookworms and roundworms seem able to avoid the attack of the system, and their sheer size makes them hard to dislodge. Pride of place among the worms seems always to go to the tapeworms because they are considerably bigger than the rest of man's parasites, and pride of place for size among the tapeworms goes to *Diphyllobothrium latum*. This is the fish tapeworm which can reach lengths of up to eight feet in man.

Infection with *Diphyllobothrium* is always a result of eating raw or partly smoked fish, usually pike. The worm is found in many parts of the world, especially around the Great Lakes, and in Alaska, Scandinavia and Japan. Larvae are taken in from an infected fish and may attach themselves anywhere in the upper intestine, where they grow into the adult worm which will produce eggs. These find their way back to the fish via fecal contamination of water, and the cycle continues.

Although the tapeworms are dramatic parasites in terms of their individual size, they are not a major health problem in worldwide terms. Perhaps the most important of the intestinal parasites is *Ascaris lumbricoides*, thought to be responsible for over thirty thousand deaths a year. Adult worms live in the small and the large intestine and may be present in extremely large numbers. It is not uncommon for individuals to carry several pounds of worms in their gut, and these worms can give rise to intestinal obstruction.

Ascaris is a parasite with a simple life cycle. The adult worms produce eggs which are passed into the feces, and ingestion of any fecally contaminated material may lead to infestation. This is especially likely to happen when night soil is used for manuring crops, a particularly common practice in some Third World countries. One reason why *Ascaris* is such a successful parasite is that its eggs are exceptionally durable; they may survive for up to seven years, even in hot conditions.

Although it is no longer a major problem in the U.S.A., the hookworm was once as big a health hazard in the southern states as *Ascaris* is in other parts of the world today. There are two major species of hookworm, *Necator americanus* and *Ankylostoma duodenale*. As the name suggests *Necator* was originally a New World parasite, while *Ankylostoma* was found in the Old World. Both parasites have a worldwide distribution in temperate and tropical regions, although *Necator* prefers the hotter, more humid, climates.

Adult hookworms live in the upper part of the small intestine by sucking blood. The mouthparts of *Necator* are adapted to attach to a single villus and then to slice off its top with razorlike plates. The parasite attaches to the villus as though it were hanging onto a teat, and the force of the blood pressure within the arteriole supplying the villus simply pumps blood through it. The parasite extracts only a small amount of food from the blood passing through it, and there is, therefore, an excessive blood loss into the intestine. Since every victim may carry thousands of hookworms, this can lead to tremendous blood loss, and it is not surprising that anemia is the major problem in many of the people affected.

Like all the other intestinal parasites, the hookworm's eggs are passed out in the feces. The female hookworm may produce as many as ten thousand eggs every day, and these hatch out on the ground into tiny larvae. They penetrate the skin of anyone coming into contact with them, and obviously they are most likely to penetrate the skin of the feet. The points where they have made their entrance can give rise to considerable irritation, which is known as the "ground itch." The problem was effectively controlled in the U.S.A by the hookworm commissions of the 1920s through the improvement of sanitation and the disposal of feces and by encouraging people to wear shoes.

Infection and infestation by parasites are the major problems that beset the small intestine; they are usually avoided by better public health and sanitation methods rather than by other forms of specific treatment. When the risk of infection is controlled, the small intestine is remarkable by virtue of the rarity of serious clinical problems arising within it.

Chapter 5

The Chemical Factory

Medicine, anatomy and physiology were, in times past, often indistinguishable from religion, superstition and divination. They have taken a time almost equal to that of the development of mankind to reach their present scientifically based state.

One organ, the liver, has always held an important place, whether in medicine or early religious practice. Babylonian priests once examined the organs of animals as an aid to predicting future events, and the liver was considered of such importance in divination that it was used only for particularly serious matters. The usual practice was to sacrifice a sheep, remove the liver and then carefully examine its surface. Models of the liver exist from this period which appear to have been a kind of reference guide for the priests. One clay model, dating from 2000 B.C., has the surface of the liver divided into fifty sections, each with a prognostication written on it; and in most of the sections there is a hole for the insertion of a wooden peg. The priests looked for abnormalities in each of the corresponding sections of the sheep's liver. This close observation of the organ was crucial in that it laid the foundations of anatomy in the further development of medicine in Greece.

Around 300 B.C., Theophilus of Chalcedon, known as the Father of Anatomy, made the first public dissections of both animals and humans. From his observations he was able to describe in unprecedented detail the liver, pancreas and biliary tree. At this time there appear early references to diseases of the liver. The great Hippocrates in his Aphorisms noted that in cases of jaundice hardening of the liver was a bad sign; perhaps here he was referring to cirrhosis, or tumors, in the liver.

Many of the early advances in medicine are associated with close observation and an interest in anatomy. In communities where dissection was not allowed, mysticism and astrology became involved. This was so in China, where the medical

Alcohol, like many of life's pleasures, is a two-edged sword. While in moderation it may liven the mood, in excess, it can cause cirrhosis, which involves irreparable damage to the functioning of the liver.

In conjunction with the liver from a dead sheep, this clay model of a liver (below) was used as long ago as 2000 B.C. by Babylonian priests to predict future events, through the position of abnormalities.

The thoughts and teachings of the physician Galen influenced medical thinking for over fifteen centuries. He dissected human corpses in defiance of Roman Law in order to develop a view of anatomy.

This figure, (right) adorned with astrological signs, was used as a guide for blood-letting — a perfect example of the bond which once existed between medicine and astrology.

Galenus

classic, the Canon of Medicine, mostly gathered together around 300 B.C., labels the liver as a military leader, the seat of anger and of the soul. The physiology of the Chinese was also incorrect, with the liver thought to be the source of tears and also of nasal secretions.

In Roman times, one of the prominent physicians, Galen (129–199 A.D.), from Pergamum in Asia Minor, came closer to appreciating the true function of the liver, for he believed that food substances absorbed from the intestines came to the liver in the portal vein as chyle. His view was that this chyle was converted to blood in the liver and that a natural spirit, pneuma, was added to help in growth and nutrition. The blood was thought constantly to be made in the liver and then used up in other tissues.

Little further was discovered of the function of the liver over the next thousand years. Astrology retained its influence: Paracelsus, known as the Luther of Medicine, who practiced in the first half of the sixteenth century, felt that dissection of the body fostered ignorance and that anatomy should be learned from astronomy. He wrote that the planet Jupiter ruled the liver, while Mars was in control of the gallbladder.

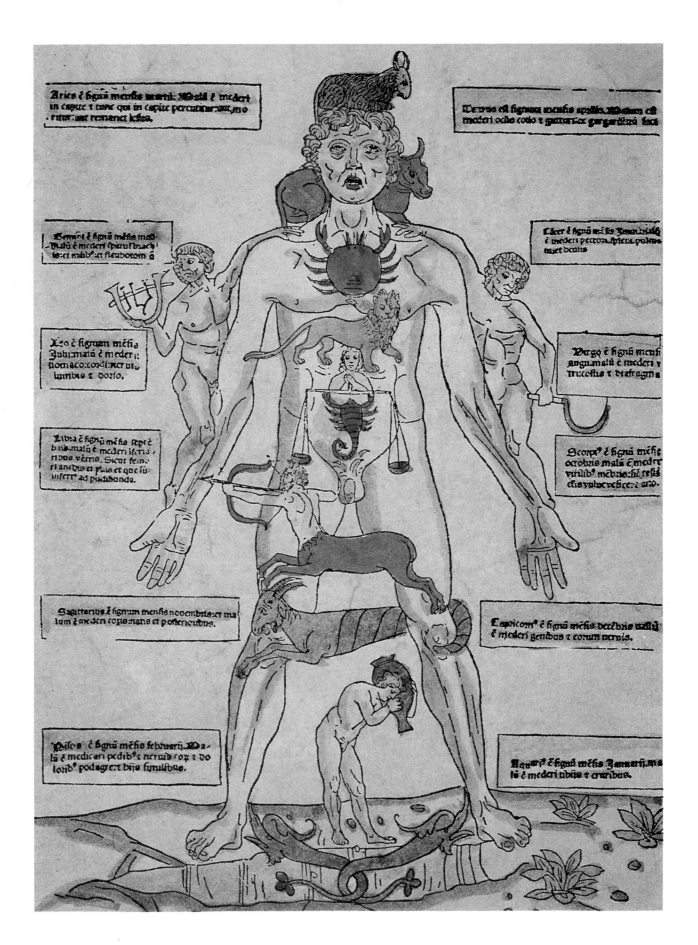

95

Unlike any other organ, the liver has two sources of blood. Eighty percent of its blood is delivered from the intestines via the portal vein and the other twenty percent is supplied by the hepatic artery.

In the sixteenth century, interest in anatomy rose again, stimulated by artists such as Leonardo da Vinci and Albrecht Dürer. The anatomist Malpighi (1628–94), a professor in Bologna, was first to show that the liver was not of uniform consistency.

It was not until the nineteenth century, however, that real advances occurred in knowledge of the structure of the organ. The great French physiologist Claude Bernard (1813–78) was, in effect, the founder of experimental medicine. In 1855, by a fortunate accident, he left a sample of liver extract longer than he intended and discovered that the liver was able to produce sugar — glucose — itself, not just to extract it from the blood. Two years later, he found that the glucose was formed from glycogen and so began our present-day understanding of carbohydrate metabolism.

The Liver — the Body's Chemical Factory

As a result of Claude Bernard's discovery of glycogen, which is stored in the liver, it soon became clear that the liver has a wide variety of functions and is chemically extremely active, which gives it an influence over all other organs. Some actions involve the breakdown of complex chemicals; other important ones involve synthesis, particularly the manufacture of protein molecules. The liver acts as a cleansing station, inactivating hormones and drugs. The Kupffer cells that line the liver's blood vessels mop up unwanted elements and infectious organisms reaching it from the gut.

In order to maintain these important functions, the liver receives a large blood supply, three pints of blood every minute. Between meals, more than three-quarters of this supply comes to the liver by way of the portal vein which drains the intestine. The remainder is from the body's main arterial system via the hepatic artery. When food is eaten, more blood is diverted to the intestine to cope with the tasks of digestion and absorption, and blood flow in the portal vein increases.

Before birth, and for a few weeks after birth, the liver is an important site for the formation of white and red blood cells. After the neonatal period, this task is taken over by the spleen and the bone marrow. Sometimes though, abnormalities of blood formation in adults may make the liver revert to its old role as a blood-forming organ.

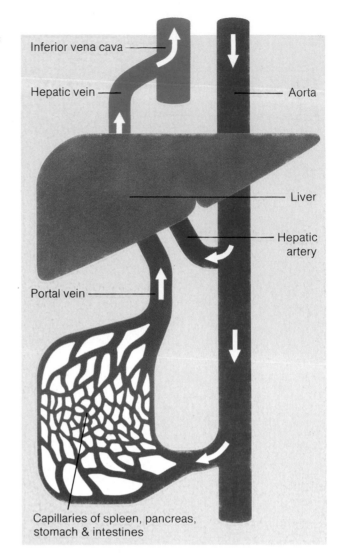

There are many substances in the bloodstream that indicate important events in the tissues of the body. These substances can be naturally occurring, or endogenous, such as the hormones produced by the endocrine glands. Alternatively they may be substances such as drugs and medicines, which are essentially exogenous, or from outside the body. In all cases, it is the task of the liver to modify these biologically active substances so that they lose their biological effect — a process known as detoxification. The fact that alcohol is largely detoxified in the liver explains why excessive alcohol consumption leads to liver disease.

When the liver has become diseased, it may lack the full capacity to detoxify both endogenous and exogenous substances. The natural hormones aldosterone and estrogen are good examples. Both these substances are present in excessive amounts when the liver is diseased for the simple reason that the liver is no longer capable of detoxifying them and switching off their activity. Normal amounts of

The liver, the largest single organ in the body, which weighs between three and four pounds in a man, is situated in the right upper quadrant of the abdominal cavity. The gallbladder lies behind the liver.

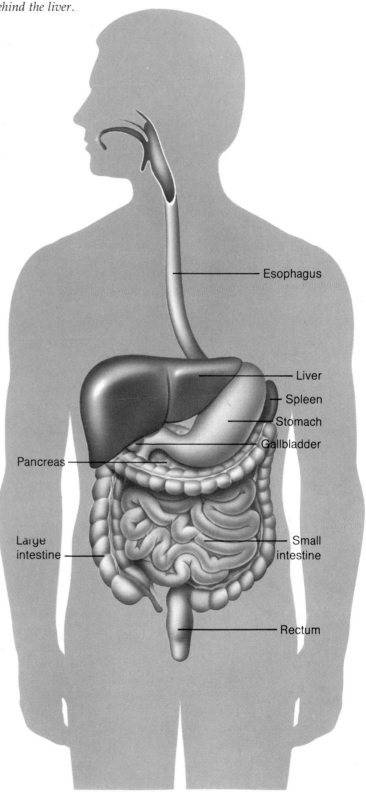

- Esophagus
- Liver
- Spleen
- Stomach
- Gallbladder
- Pancreas
- Large intestine
- Small intestine
- Rectum

aldosterone in the blood are responsible for the correct degree of salt and water retention; when not detoxified, as occurs in chronic liver disease, its excessive activity causes tissue swelling and water retention. An excess of estrogen — the female hormone — leads to the development of female characteristics, and men with chronic liver disease start to develop breast tissue — gynecomastia.

Drugs will also tend to have longer-lasting effects when the failing liver is unequal to the task of detoxification, and this means that all drugs must be prescribed with special care if a patient is known to have chronic liver problems.

However, the liver's importance does not end at the detoxification of hormones and drugs. It is, of course, essential to the way in which the body handles nutrients — all the products of digestion are brought to it. The liver builds up stores of glycogen from glucose absorbed by the intestine, with excess glucose being carried to the body's adipose cells to be laid down as fat. Fat carried to the organ from the intestines can be used there for various purposes and especially for the manufacture of cholesterol, a substance needed by the body for hormone manufacture, which is then carried into the bloodstream bound up with protein in the form of lipoprotein.

The liver is also central to the way in which the body handles protein. The blood contains large amounts of different proteins, many of which are made in the liver. However, it is doing more than simply reconstructing proteins from nutrient amino acids absorbed via the intestine; it also breaks down unwanted amino acids and converts them into amino acids which the body may need

The liver is, therefore, the major site in the body for the chemical handling of proteins. Proteins are distinguished from carbohydrates and fats by the fact that they contain nitrogen in addition to the carbon, hydrogen and oxygen of the other common compounds. In the course of its rearrangement of proteins and amino acids, the liver produces a certain amount of free nitrogen, normally in the form of a toxic substance, ammonia. This is rapidly converted into urea — the body's main waste product from the liver's chemical juggling of proteins — which then enters the bloodstream, to be excreted by the kidneys.

The basic functioning units of the organ are its cells, or hepatocytes, which are arranged in a highly ordered way. The basic structure of the liver is critical to its function, and any disorder of the structure can lead to disturbances in its working.

The liver is the largest organ in the body and weighs between three and four pounds in men and rather less in women. At birth, at about five ounces, it is even larger compared to the total body size: one twenty-fifth of total body weight. It lies mainly in the right upper quarter of the abdomen, under the diaphragm. There are two main lobes, the right lobe and a smaller left one which crosses the midline to lie above the stomach. Against the undersurface of the liver, which is almost completely covered by peritoneum, the lining layer of the abdominal cavity, lie the stomach, duodenum, colon, right kidney and the gallbladder.

These two lobes are further divided into lobules at the center of which are small blood vessels, draining into the hepatic vein. This takes blood into

the inferior vena cava, the major vein draining blood from the lower half of the body. Between the lobules lie the portal canals which contain three structures. First, there are fine branches of the hepatic artery and portal vein; blood from these flows through the lobule to reach the second, central structure, the tributary of the hepatic vein. Finally, there is a channel to collect bile, which flows out of the lobule in the opposite direction.

Alcohol and the Liver

Countries such as the U.S.A. and France, where alcohol consumption is high, have high rates of alcoholic liver disease; where less alcohol is drunk, for instance in Finland, such liver disease is much less common. It does not seem to matter what sort of alcohol is drunk; beer, whiskey, gin, wine, it is all much the same to the liver which metabolizes ninety-five percent of the alcohol that is consumed. At any given level of alcohol intake women are more prone to liver problems than men.

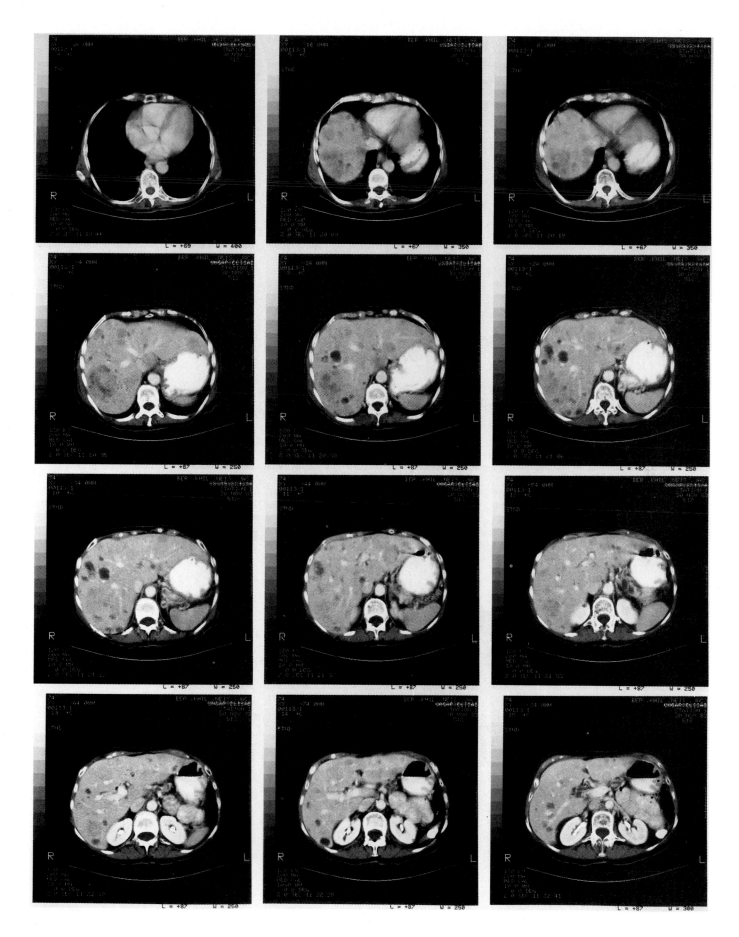

The liver is the body's main chemical processing station. It is particularly concerned with handling nutrients from the intestine, hence the fact that most of its blood supply arriving in the portal vein has already passed SUPERIOR

through the intestine. It is also heavily involved in dealing with toxic substances both from without and within the body. The structure and function of the liver are closely related and any destruction of its

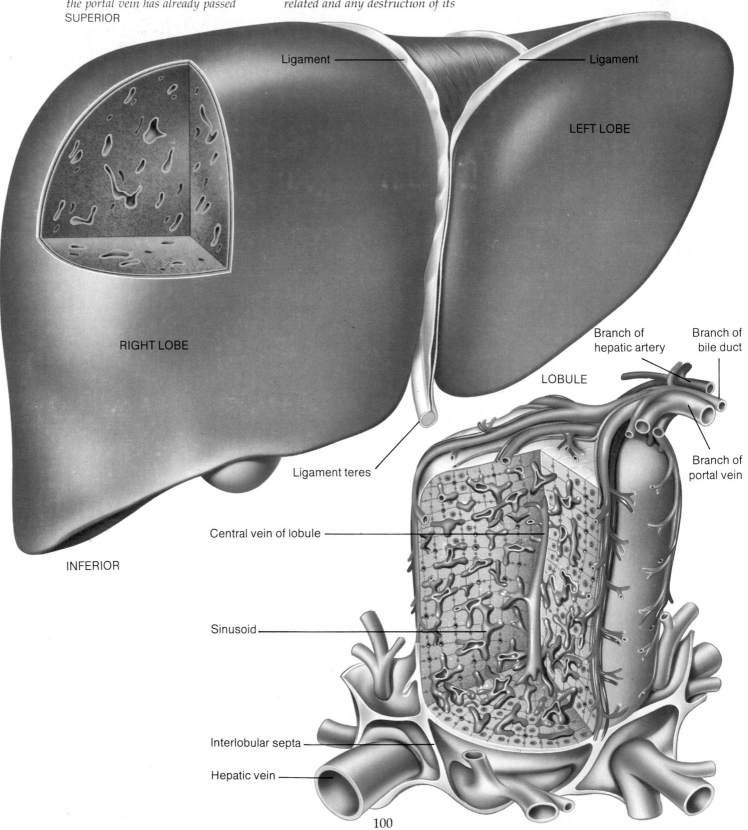

Ligament

Ligament

LEFT LOBE

RIGHT LOBE

Branch of hepatic artery

Branch of bile duct

LOBULE

Branch of portal vein

Ligament teres

Central vein of lobule

Sinusoid

INFERIOR

Interlobular septa

Hepatic vein

100

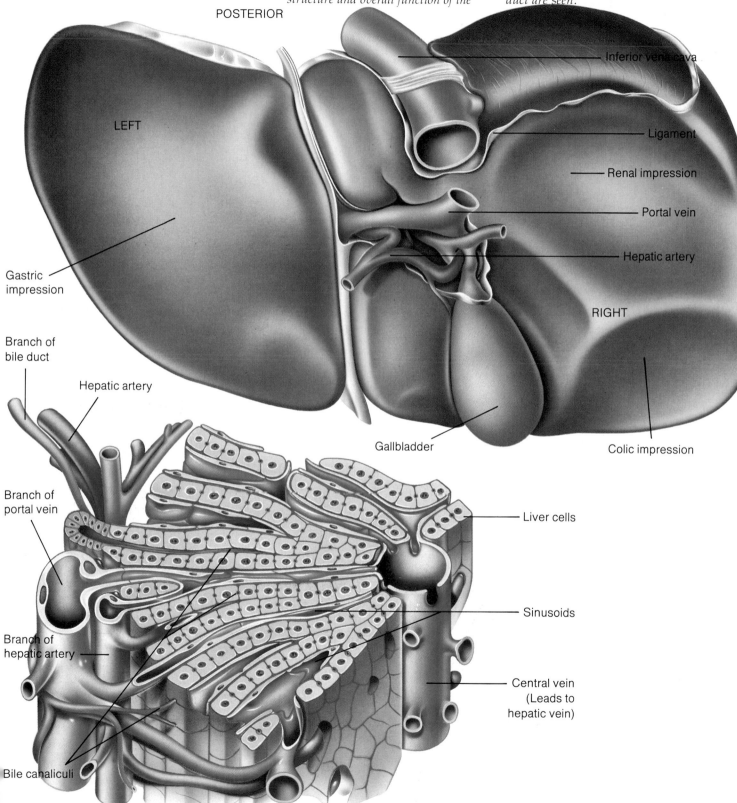

complex microarchitecture, as
happens in cirrhosis, disrupts its
function. Although the organ is
divided into a larger right and a
smaller left lobe, the internal
structure and overall function of the

two lobes is the same. Visualized
from the front (left), the surface of
the liver is relatively smooth. Viewed
from below (right), the major blood
vessels and the gallbladder and bile
duct are seen.

POSTERIOR

LEFT

Inferior vena cava

Ligament

Renal impression

Portal vein

Hepatic artery

Gastric
impression

RIGHT

Branch of
bile duct

Hepatic artery

Gallbladder

Colic impression

Branch of
portal vein

Liver cells

Sinusoids

Branch of
hepatic artery

Central vein
(Leads to
hepatic vein)

Bile canaliculi

101

The mildest form of damage produced by alcohol is an increased amount of fat in the cells of the liver. A fatty liver looks large and yellow, but this is not necessarily a permanent change, for if alcohol intake ceases, the organ, thanks to its regenerative powers, is able to return to normal.

The next stage of liver damage by alcohol is acute alcoholic hepatitis. Hepatitis means inflammation of the liver cells, and alcohol can cause inflammation and even cell death. In this stage, the liver is often large and tender, and the patient feels ill and feverish and is often jaundiced. Occasionally, in the condition known as hepatic encephalopathy, the liver may fail completely and coma occurs.

Acute alcoholic hepatitis may progress to cirrhosis. This may develop quietly, without any particular signs of ill health, but in the end the yellow skin color typical of jaundice is obvious, and all the functions of the liver begin to fail. These changes are not inevitable in heavy drinkers — people who consume more than one-third of a bottle of spirits per day — for only about one in ten of those who drink heavily for years will go on to scar their liver with alcoholic cirrhosis.

Cirrhosis of the Liver

In cirrhosis there is a fine network of scar tissue — fibrosis — throughout the structure of the liver. This widespread fine scarring divides up the cells into small islands, or nodules, in which there is regrowth and repair of tissue. The liver has an amazing capacity to repair itself after all types of damage, but the cirrhotic fibrosis leads to permanent disruption of the structure, which results in a major loss of function.

A common complication of cirrhosis is a rise in pressure in the portal vein bringing blood from the intestines. The nodules in the liver obstruct the tributaries of the hepatic vein and increase the resistance to blood flow and thus raise the portal vein pressure. This development of portal hypertension has severe repercussions for patients.

In portal hypertension it is difficult for blood to flow through the liver, so other paths, called collateral vessels, are opened up which allow blood from the intestines to return to the heart by short-circuiting the liver. The collateral vessels are wide, thin walled veins and appear either in the lining of the esophagus, where they are called esophageal varices, or in the rectum as large hemorrhoids. Occasionally veins on the front of the abdomen become large, spreading out from the umbilicus. This is called a caput Medusae, since the veins are thought to resemble the head of Medusa, one of the three Gorgons in Greek mythology, who had snakes for hair. The additional pressure also causes the spleen to enlarge.

The danger from all these enlarging veins is that those in the esophagus may rupture. When this happens, the bleeding is often dramatic and difficult to stop; and the mortality rate of bleeding from esophageal varices is extremely high.

Doctors may detect other signs of chronic liver disease when diagnosing cirrhosis. There may be jaundice, the palms of the hands become red, the nails white and opaque; in males the testes become small and in women breasts enlarge. On the skin little dilated blood vessels may be seen, they are likened to tiny spiders with small red bodies and spindly red legs and are called spider nevi.

Although the liver is usually small and shrunken in cirrhosis, the abdomen often becomes enlarged because of a build up of fluid, a condition known as ascites. Patients who suffer from cirrhosis are susceptible to infection, and tuberculosis may be a particular problem.

Other Liver Problems

Alcohol is not the only cause of cirrhosis of the liver, although it is certainly the most common in the U.S.A. There are many other chronic diseases which seem to attack the liver tissue and cause the same kind of scarring as happens with alcohol.

The only way a physician can be certain that the liver is sufficiently scarred to be called cirrhotic is to examine a specimen under the microscope. Fortunately, it is no longer necessary to perform an abdominal operation to obtain such specimens; with the aid of a special biopsy needle, they can be obtained through the skin.

The technique of liver biopsy has added greatly to medical knowledge of cirrhosis, especially the forms that are not due to alcohol. It appears that, in many cases, cirrhosis results from the body's own immune defense system turning against the normal liver cells and triggering the inflammation that leads to the condition. Exactly why this should happen remains a mystery, but there may often be some sort of triggering event, such as infection of the liver by a virus. There are many different viruses which can attack the liver, and in recent years much research has centered around the activities of the hepatitis B virus.

Many viruses are capable of producing the inflammation of the liver known as hepatitis. The commonest is called hepatitis A virus, but there are others, such as hepatitis B and the oddly named non-A, non-B hepatitis virus. Approximately one-half of all Americans have blood which, when tested, indicates a previous infection with one of these viruses. Since nowhere near this number of people have actually had jaundice, this shows that on many occasions the illness must be extremely mild and not recognizable as hepatitis. Much less common is hepatitis caused by other viruses, such as infectious mononucleosis (glandular fever), cytomegalovirus, herpes simplex and yellow fever.

Hepatitis A virus is a tiny RNA virus which is usually passed onto new sufferers by contaminated food or water. Oddly, cases are most common in the fall, and children and young adults are most

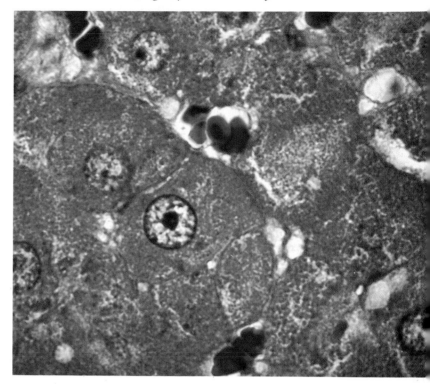

The liver cells — hepatocytes — have the ability to regenerate. Following their destruction and subsequent regeneration as happens in certain forms of hepatitis, the liver will again function normally.

often infected. There is an incubation period of two to seven weeks between being in contact with the virus and developing the illness. Just before and as the symptoms start, sufferers excrete great numbers of the virus in their feces, and they are most likely to pass on the disease at this time, often before they realize that they are ill. Fortunately there are no long term problems from hepatitis A infection; patients do not develop chronic liver disease, nor do they become chronic carriers.

Hepatitis B virus, often called serum hepatitis, is a DNA virus larger than hepatitis A. Transfer from one person to another is generally by blood products or secretions. Saliva, tears, semen, vaginal secretions and breast milk have all been found to contain the virus. Blood transfusion and use of contaminated injection needles have often been shown to be responsible for transmission of the disease. Hospital workers and patients on hemodialysis or receiving numerous blood transfusions are particularly likely to be infected in this way. The virus can be passed on during kissing and sexual activity. Infection rates are high among female prostitutes and male homosexuals.

103

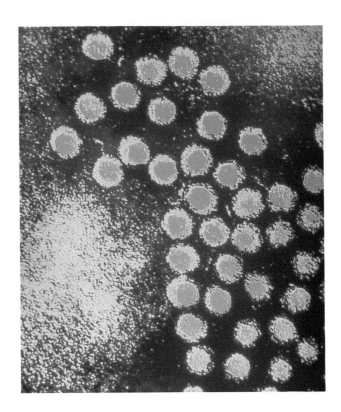

Generally, hepatitis A virus — a tiny RNA virus — (right) is milder in its effect than hepatitis B virus — a larger DNA virus — (far right) which kills one in every hundred sufferers as a result of liver failure.

The problem with hepatitis B infection is that, unlike hepatitis A, once infected, the patient may carry the virus and be infective for ever. Ten to fifteen percent of those infected will do this, and about two percent will go on to develop a chronic form of hepatitis and cirrhosis. The incubation period for hepatitis B is longer than for A, with a gap of from seven to twenty-five weeks from first contact with the virus before the trouble starts.

When blood tests became available to detect hepatitis A and hepatitis B, it was found that some patients had a similar disease without evidence of either virus. This is how the non-A, non-B hepatitis virus acquired its bizarre name.

The acute hepatitis produced by all three viruses is remarkably similar, although hepatitis A is usually milder. For up to a week, the patient feels unwell, tired, sick, and may have joint pains; often an intense dislike of fatty foods will develop. After about a week, the patient feels better, but the yellow color of jaundice is seen and the liver may be tender to touch. Jaundice is seen most easily in the whites of the eyes and the palms of the hands.

Blood tests will show antibodies against the virus and increased bilirubin in the blood; they will also show a rise in liver enzymes, such as transaminases, which leak out of the inflamed liver cells infected by the virus. Unfortunately one in a hundred patients with hepatitis B infection dies of acute liver failure because the virus destroys so many of the liver cells. Also, some unlucky hepatitis B sufferers remain chronic carriers of the virus or are left with chronic liver disease.

Using the techniques of genetic engineering, it has been possible to manufacture a vaccine which can provide protection against hepatitis B. By manipulating the genetic material of bacteria it is possible to make them manufacture a protein which is identical with the protein covering of the hepatitis virus. This will trigger the body's immune defense mechanism without bringing about the damage that the whole virus would cause. At risk individuals, such as male homosexuals or health care workers, can, therefore, be given the vaccine so that if they do become infected their defense mechanisms will be ready to combat the virus and will succeed in eliminating it.

Although immunization can be used to avoid the disease if contact is likely, there is little that can be done once symptoms have developed. No drugs are of any use, but it is best for a patient to rest in bed for a week or so in the early stages and to avoid alcohol until the liver enzyme tests return to normal — this may take months.

Drugs and the Liver

The liver — the great detoxifier — is crucial for the breakdown of numerous drugs. This means that when the organ is not acting properly, drugs may build up in the body and produce unexpected toxic effects, even at normal doses. It also means that the liver is at risk of damage from the breakdown products of the drugs on which it acts.

Drugs can also stimulate the enzyme systems of the liver, making them more active. When this happens, the tiny smooth membranes of the liver cells grow larger. Barbiturates, alcohol and antiepileptic drugs are the best known compounds inducing the liver enzymes in this way.

Acetaminophen is a very effective painkiller, perfectly safe in the usual doses given. Even at normal dosage levels, however, acetaminophen is changed in the liver cells to a potentially dangerous by-product, which is made safe by combination with a substance in the liver called glutathione. When an overdose of acetaminophen is taken, there is insufficient glutathione available in the liver, and its cells are damaged. This can be avoided if suitable

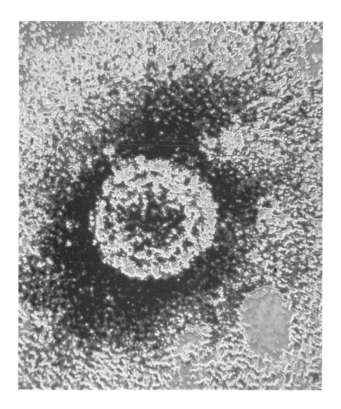

antidotes can be given to the patient within ten hours of taking the overdose.

Occasionally — and unpredictably — other drugs can cause liver problems. These can occur with anesthetic agents, such as halothane, and with anti-tuberculous drugs such as rifampin and isoniazid.

Liver Failure

From time to time the liver cells are subjected to an experience that destroys so many of them that an insufficient number remains to carry out the organ's various important and vital functions; this is known technically as liver failure.

Because the liver is active in so many processes, its failure causes a number of major disturbances in body function. Many cases of acute (sudden) liver failure are a result of viral hepatitis, although only about one percent of such infections ends with liver failure. More common is failure occurring on top of chronic (long term) liver diseases, precipitated by an infection of a gastrointestinal hemorrhage, or a drug, or alcohol.

Liver failure causes an accumulation of those toxic products usually broken down in the organ. These products, including ammonia, interfere with the function of the brain, and the patient becomes drowsy, then lapses into a coma. While drowsy, the sufferer will become confused and may develop a shaking of the hands, which is often known as the flapping tremor.

The sugar level in the blood falls as glucose is no longer produced. Infection is common, and the kidneys begin to fail; as a consequence of this, fluid is retained in the body. The output of blood-clotting substances usually produced by the liver decreases so that uncontrolled bleeding becomes increasingly likely and is often a cause of death. If a coma develops, most such patients will die.

The powers of recovery of the liver are, however, remarkable. In fact, if a patient is lucky enough to recover from acute liver failure caused by viral hepatitis, there will usually be no long term problems, even though most of the liver cells have been destroyed during the illness. The important difference between sudden liver failure and cirrhosis is that the basic structure of the liver is not disturbed in the former. This means that when the cells regrow, they do so in an ordered fashion which permits normal function.

In cases of chronic liver failure, it is now possible to carry out a transplant. Much of the pioneering work in this field has been performed by the Pittsburgh based surgeon, Thomas Starzyl. Apart from the technical aspects of joining all the blood vessels and bile ducts together, the main problem, as in other transplants, is rejection. The body's guarding immune system recognizes that the new liver does not belong and tries to destroy it; in addition, the drugs used to suppress this response cause problems of their own. Nevertheless, liver

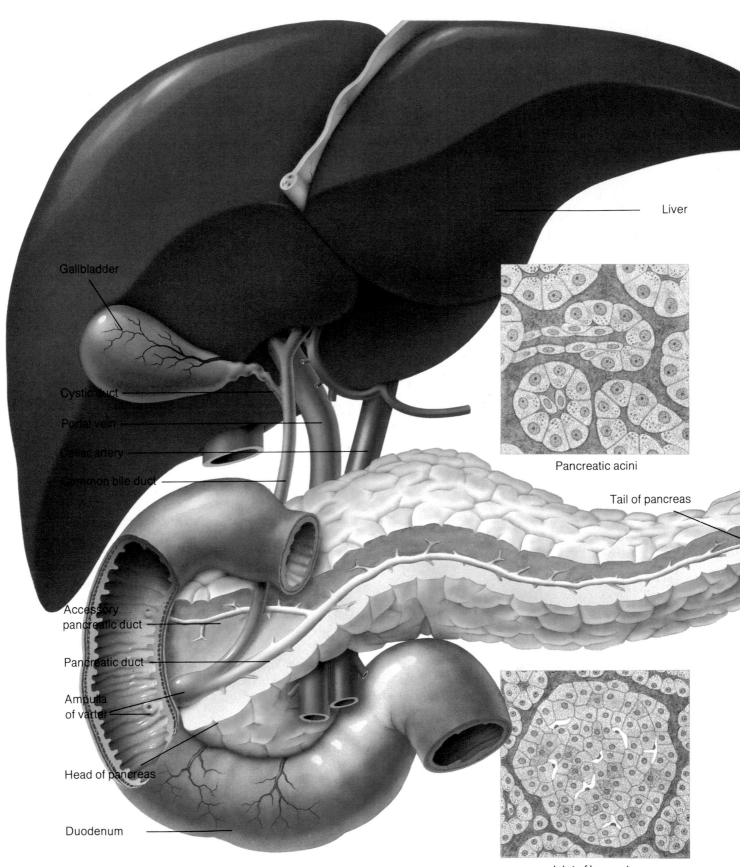

Liver

Gallbladder

Cystic duct

Portal vein

Celiac artery

Common bile duct

Accessory pancreatic duct

Pancreatic duct

Ampulla of vatter

Head of pancreas

Duodenum

Pancreatic acini

Tail of pancreas

Islet of Langerhans

transplantation offers a hope for patients with an otherwise untreatable disease and has a one year survival rate of about fifty percent.

The Gallbladder

Bile is a substance that has long fascinated humankind. The great Pythagoras, better known for his mathematical prowess, established a theory of life based on four elements: earth, air, fire and water. These formed the four humors of the body; blood, which was hot and moist, yellow bile, which was hot and dry; phlegm, which was cold and moist, and black bile, which was cold and dry. The relative proportions of these four humors were thought to determine health and intelligence. Remnants of this philosophy lasted from Pythagoras' time, around 500 B.C., until the mid-nineteenth century, when German pathologist Rudolph Virchow banished the last cobwebs of Pythagoras' and Galen's medical theories. However, our language today still echoes Pythagoras' theories, with terms such as melancholy, derived from melancholia, or black bile.

The fluid is formed in the bile canaliculi which lie between the cells in the lobules of the liver. It is collected in the branches of the hepatic duct lying in the portal canals between these lobules. These small hepatic ducts join to produce the right and left hepatic ducts that lead out from the two lobes of the liver and combine to make the common hepatic duct, through which passes all the bile produced.

The common hepatic duct runs for one and a half inches before being joined by the duct coming from the gallbladder, the cystic duct. The combination of common hepatic duct and cystic duct produces the common bile duct, which passes down behind the duodenum and through the head of the pancreas to open into the duodenum. Just before it enters the duodenum, it is joined by the main pancreatic duct, carrying all the digestive juices from the pancreas.

The small protrusion where the bile duct enters the duodenum is known as the ampulla of Vater. It is surrounded by a specialized piece of muscle, the sphincter of Oddi, which is able to control the opening and, therefore, the flow of bile and pancreatic juices into the intestines.

The gallbladder is a pear shaped pouch that protrudes from the bile duct. There is a great deal

of variation, but a normal gallbladder generally holds about two fluid ounces of bile. It concentrates the bile from the liver and holds it ready for discharge into the duodenum. The gallbladder narrows into the cystic duct by a necklike structure, Hartmann's pouch, and it is at this narrowed area that gallstones often tend to stick. If they pass this area, they may lodge in the common bile duct or, if they are small enough, pass right through it into the duodenum. In the wall of the gallbladder there is a smooth muscle which allows it to squeeze out the bile when this is needed for digestion.

The Function of Bile

The Swiss are renowned for their diligence, and Albrecht von Haller was typical in this respect. The breadth of his learning was truly breathtaking, embracing poetry, botany and medicine, as well as the classics and theology. When the new University of Göttingen was looking for its inaugural professors in 1736, Haller's fame was such that he was given

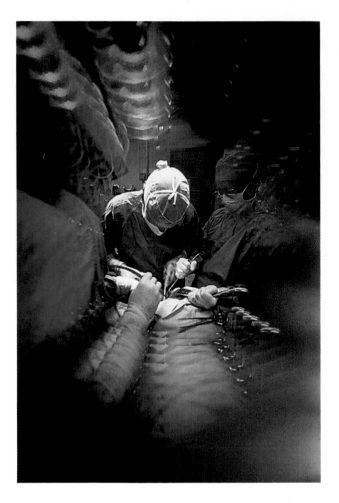

Operations on the gallbladder once had a mortality rate of thirty percent. Today gallbladder surgery is considered routine. This operation (right) was photographed through a zoom lens.

the chairs of Botany, Medicine and Anatomy, despite the fact that he was still only twenty-eight years old.

Haller made many contributions to the field of medicine and, in 1736, published his observations on the functions of bile, pointing out that it was essential for the digestion of fats. Over a century later, Theodor Schwann, the German physiologist who characterized the activity of the enzyme pepsin, followed up Haller's observations and showed that bile emulsified fat before it could be digested.

Bile contains a mixture of bile salts, phospholipids, cholesterol, pigments, proteins and inorganic ions such as sodium. Its yellow color is caused by bilirubin, mainly formed from the breakdown of red blood cells that have reached the end of their normal life span of about four months. A bilirubin build up is responsible for jaundice — yellow coloring of the skin — when the liver is diseased or the flow of bile obstructed. Approximately two and a quarter pints of bile are produced by the liver every day. This is collected up in the small biliary vessels in the liver and passes to the gallbladder where it is stored. However, the gallbladder is more than just a storage vessel; in it there is such great active reabsorption of salt and water that the volume of bile is reduced to only one-tenth of its original quantity. It changes from a thin liquid to a thick, mucuslike material. The bile salts are not absorbed, so their concentration is increased by about ten times.

The excretion of bile is controlled by a hormone called cholecystokinin, which is released into the blood by the duodenum when food enters it from the stomach. The cholecystokinin travels in the blood to the gallbladder and triggers contraction, causing the concentrated bile to be ejected.

The bile salts are essential for the emulsification of fat globules in the chyle. Once they have been involved in carrying emulsified fat from the lumen of the intestine to its epithelial surface for absorption, they travel on down the intestine. They are not, however, lost to the body in the feces, for the last part of the ileum reabsorbs most of the bile salts that enter it. Once reabsorbed by the terminal ileum, the bile salts are carried back to the liver to be collected again in the bile and secreted into the duodenum. This system is known as the enterohepatic circulation of bile salts. It works so efficiently that the entire bile salt pool of the body may pass through the system twice or more during the absorption of a single meal.

Gallstones

Much of the learning which finally released European medicine from the iron conservatism that had bound it since the time of Galen was based at the two Italian universities of Padua and Bologna. At both universities, the anatomical schools dissected human bodies, and the anatomists concentrated on rational observation of facts, rather than on a reliance on the delivered "wisdom" of historical authorities. Without this iconoclastic challenge to the word of Galen, medicine would probably have languished in an intellectual desert long after the Dark Ages and the Renaissance.

One of the Paduan anatomists, Gentile da Foligno, was the first to observe gallstones, although he did not attribute any special significance to them. In those days, the gallbladder was believed to be the seat of courage within the body; and black and yellow bile were regarded as two of the four humors.

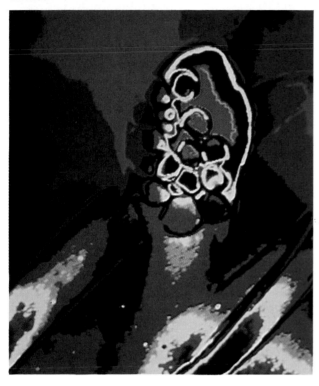

salts or pigments such as bilirubin. Cholesterol is the commonest constituent, but many of the stones are of mixed composition. In normal circumstances, cholesterol is prevented from flowing out of the body in the bile by a delicate balance of bile salts and phospholipids. A change in this balance can produce an increase in cholesterol and a decrease in bile salts, and this in turn can lead to formation of gallstones.

There may be a single stone or a whole collection of stones in the gallbladder. Often they remain there for many years, causing no distress at all. But problems, usually either inflammation or obstruction of the common bile duct, can arise in about a quarter of cases.

A plain X ray of the abdomen will show up gallstones in only fifteen percent of cases, because only this number contain enough calcium to be visible on an X-ray film, and usually a more sophisticated investigation will be necessary. This may be by the use of an X-ray dye which concentrates in the gallbladder and shows up on the X ray; gallstones then appear as filling defects in the gallbladder. The dye may be given by mouth or directly into a vein.

A simpler technique, which seems to be just as reliable, is ultrasound, first used to detect submarines in wartime. Sound waves at frequencies above those heard by the human ear are transmitted, and the returning echoes monitored; like submarines in the sea, solid gallstones show up well in liquid bile.

If a gallstone does slip out of the gallbladder and into the biliary tree — assuming that it is too large to pass easily through the bile duct and out into the duodenum — it will cause obstruction of the biliary tree, an extremely serious problem. Biliary obstruction due to stones is one of the common causes of jaundice. Because bile cannot pass out of the biliary tree into the duodenum, it backs up, with the result that the bile cannaliculi become blocked. This in turn means that the liver cells cannot excrete bile, and there is a build up both of bile and the substances from which it is made in the bloodstream. As bilirubin begins to build up, it stains the skin, so producing the yellowing which characterizes jaundice.

When a patient goes to a physician with jaundice, the first question in the physician's mind is whether

Along with so many of his contemporaries, Gentile da Foligno suffered a premature demise in 1348, as a result of the Black Death.

So, although gallstones were discovered fairly early in the history of medicine, it was not until the latter half of the nineteenth century that advances in anesthesia allowed successful operations to remove such stones. In 1867, an Indianapolis surgeon, J.S. Bobbs, reported one of the first removals, opening up the gallbladder, removing the stone and closing the gallbladder again — a cholecystotomy. Then, in 1882, Langenbuch successfully removed a gallbladder containing stones, an operation called cholecystectomy. Four years later, Justus Ohage in St Paul, Minnesota, performed the first cholecystectomy in America. At first the operation was not popular, partly because of a mortality rate of thirty percent, but as techniques improved, it gradually became an established part of modern surgical practice.

Gallstones are present in the gallbladders of about one in five women in their late fifties, but in most of these they will not cause problems. Gallstones may be made of cholesterol or calcium

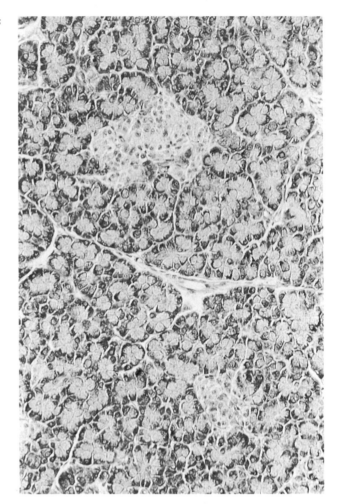

the jaundice is due to an obstructive cause or whether it results from some disease of the liver cells — such as hepatitis. Fortunately, there are now many tests available to help resolve this diagnostic problem. One of the first, and one of the most convenient, is to scan the liver using ultrasound. The ultrasound beam will not only reveal gallstones, but it will also show whether the small bile ducts in the liver have become dilated as a result of obstruction.

Alternatively there are a number of radioisotope tests available which pick up obstruction, and the use of special scanning tracers, such as technetium labeled iminodiacetic acid, can show that the cystic duct is blocked.

In the past, surgery was the only answer to biliary obstruction, but there are now two new techniques, both of which can relieve obstruction without recourse to the surgeon's knife. This not only avoids the inconvenience of an operation but spares the often extremely ill patient with obstructive jaundice from the hazards of surgery.

The first of these new techniques relates to the removal of stones. In the hands of a skilled practitioner, it is possible to identify the ampulla of Vater via an endoscope positioned within the duodenum. A fine tube can be inserted into the bile duct by this means, and X-ray dye can be injected to outline the bile duct and so highlight any stone which is present. Even more than this, a skilled endoscopist may be able to "fish out" the stone with the aid of a special miniature basket passed down the endoscope. In the same way, the stone can be made to drop out of the bile duct into the ampulla by widening the neck of the ampulla, using a special cutting wire.

The second technique is performed by radiologists, who also have an amazing repertoire of techniques which can help in cases of obstructive jaundice, especially when the obstruction is not due to a stone but to a tumor blocking the bile duct. While the endoscopist can outline the obstruction by injecting opaque X-ray dye from below, the radiologist's approach is to inject the dye above the obstruction.

A fine needle is passed through the right side of the body, level with the lower ribs. The needle passes into the liver and, by careful positioning, it may come to lie in a bile duct. By injecting X-ray dye, X-ray pictures of the biliary tree can then be taken. However, the story does not end there, for it is also possible to pass a fine wire through the needle, and, once this wire is in place, it can actually be used to thread a passage around the obstruction and down into the duodenum. A tube is then threaded over the wire and pushed into position so that it lies within the bile duct and forms a bypass to the obstruction. This technique of inserting what is known as an endoprosthesis can save an unnecessary operation, particularly in patients with incurable tumors.

The Pancreas

Early beliefs were that the pancreas might be a cushion for the stomach, or that it acted as a gallbladder to the spleen. In 1642, Johann Wirsung, working in Padua, dissected the pancreatic duct, the first such exocrine gland duct to be identified. Thirty years later, Reiner de Graaf, in Holland, used the quill of a wild duck to drain the pancreatic duct and collect the output, so establishing the

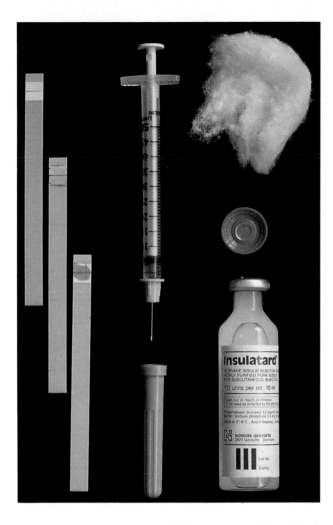

were often abnormal. The brilliant culmination of this work came in 1921, when, in Toronto, Frederick Grant Banting and Charles Herbert Best isolated insulin, a hormone whose secretion is essential to the regulation of glucose metabolism.

The pancreas lies outside the peritoneal cavity behind the stomach. It sits on top of the large blood vessels, the aorta and the inferior vena cava. It is a long, soft glandular structure, angled obliquely across the abdomen, and has a bulky head portion which fits neatly into the C-shaped duodenum. There is a slight narrowing at the neck of the pancreas and then a gradual decrease in size through the body to the tail of the organ. The tail reaches right up to the spleen, at the top of the abdomen.

There is one main duct, running right along the pancreas, to carry its glandular secretions into the duodenum. This main duct joins the common bile duct just before entering the duodenum at the ampulla of Vater. A second, smaller accessory pancreatic duct enters the duodenum just above the main duct. The pancreatic secretions are produced in lobules, each with a tiny duct communicating eventually with the large pancreatic ducts.

In between these lobules are the islets of Langerhans, the source of insulin, which is released directly into the blood and not into the pancreatic ducts. These islets make up less than one-hundredth of the pancreas by weight.

The vagus nerve, which has some action at most levels of the gastrointestinal tract, is able to stimulate the pancreas to produce secretions, and this nervous regulation induces a rapid response at the start of eating. The secretions, three to six pints of which are produced every day, have two very distinct elements.

First, when acid leaves the stomach and enters the duodenum it releases secretin from the mucosa. This secretin acts on the pancreas to produce a watery juice, rich in bicarbonate, which is an ideal neutralizing agent. Second, when food leaves the stomach to enter the duodenum, it releases cholecystokinin from the mucosa. As well as making the gallbladder contract, cholecystokinin produces a pancreatic juice rich in the essential enzymes needed to digest food; and the alkalinity of the juice produces exactly the right environment for the enzymes to reach a peak of activity.

active secretory function of the pancreas. Details of the true function of the pancreatic juice began to emerge toward the end of the nineteenth century, when Richard Kühne the Austrian chemist was able to isolate trypsin, a major active enzyme constituent of the pancreatic juice.

In the seventeenth century, evidence for another function of the pancreas began to emerge. Johann Conrad Brunner, on removing the pancreas from dogs, noticed that the animals became very thirsty and passed large volumes of urine. These are characteristic features of one type of diabetes, but the association was not made at the time.

In 1831, Richard Bright, a physician at Guy's Hospital in London, noted that there were often pathological changes in the pancreas in cases of diabetes. Then, in 1869, a German student of medicine Paul Langerhans described the islandlike structure of cells which bear his name and are the actual source of insulin; these are scattered throughout the pancreas. Thirty years later, Lindsay Opie at Johns Hopkins Hospital showed that in diabetes it was these islets of Langerhans which

Chapter 6

The Journey Ends

Many people seem to have a morbid preoccupation with the efficiency, or otherwise, with which their bowels evacuate the waste material that is left after digestion. In many societies the movement of the bowels is even taken as one of the prime indicators of the state of health. In an earlier chapter, the focus was on the small intestine, but now attention moves farther down the long path taken by food, and the spotlight is on the colon — responsible for the last processing stages and, finally, for the elimination of waste.

Digestion of the contents of the small intestine takes place almost continuously, and these contents are constantly discharged into the first part of the colon. Even when there is no food in the small intestine, secretions and matter, such as lining cells from the epithelium, pass into the colon; if it were not adapted to act as the store for all this waste material, defecation would be virtually continual.

Apart from storage, the colon has another important function — the removal of the water that finds its way into the food residue as a result of the secretions of the stomach and intestines. Without the colon to reabsorb water from feces, there would be a considerable fluid loss, and we would have to drink a great deal more liquid than we normally do in order to avoid dehydration.

The Appendix

Although the colon performs these two vital functions, there is a particularly troublesome appendage attached to the first part of it that has no known function — the appendix.

This organ is found only in humans, in certain of the great apes and in the wombat. We can, therefore, consider ourselves unlucky in this respect, since it would be hard to imagine a more useless organ capable of giving so much trouble. In fact, appendicitis is the commonest cause of emergency surgery in the Western world today.

The full name of the appendix is the vermiform appendix — vermiform means wormlike, and the

The figs being gathered in the painting (left) have long been thought to be a particularly effective aid to regular bowel movement. Syrup of figs is still often given to children as a laxative.

Like the rest of the intestine, the appendix contains bacteria which in normal circumstances are harmless. If, however, the entrance to the appendix becomes blocked, this can lead to an increase in the number of bacteria, and the appendix becomes inflamed. At this point, the sufferer may feel a dull pain in the lower right part of the abdomen as the appendix swells and fills with pus. If this goes untreated, the appendix will eventually burst, and the pus will discharge into the abdomen. If, as sometimes happens, the omentum envelops the appendix and seals off the area, the infection is localized and an appendix abcess results. This

adjective accurately describes its appearance, for it is long and circular in cross section, like an earthworm. Its diameter is constant, about half an inch, but its length can vary from one to four inches; it is attached to the cecum and has a hollow lumen which connects it to the lumen of the large bowel. The appendix is made of muscle, beneath which is a layer of lymph tissue, and it is lined by epithelium resembling that found lining the colon. It is similar to a narrow piece of bowel with a blind end and, like the bowel, is capable of peristalsis. It receives its blood supply from a branch of the artery that supplies the ileum but also has a fold of peritoneum attached to it known as the bloodless fold of Treves. (Henry Treves was a nineteenth-century surgeon at the London Hospital, Whitechapel in the city of London, England, who is perhaps best known for having befriended the deformed gentleman known as The Elephant Man.)

There are only two diseases which affect the appendix. The first is a malignant tumor, which can affect either the glands in the lining of the appendix or can arise from special cells in the wall of the appendix. Both of these tumors are extremely rare.

The second disease is a common condition in the Western world, acute appendicitis or inflammation of the appendix. Strangely, it is a relatively new disease, and appendectomy, the operation to remove the appendix, became acceptable only after the advent of general anesthesia in the middle of the last century. In fact, despite this advance the operation was not widely practiced until the end of the nineteenth century; and even well into this century, there were many physicians and surgeons who did not believe that this was the best form of treatment for acute appendicitis.

One of the most famous cases was that of King Edward VII. In 1901, at the age of fifty-nine, he developed acute appendicitis a few days before he was to be crowned and had an operation the evening before his coronation should have taken place. There was considerable argument among the royal physicians about whether or not to operate, and also as to whether the King's appendix should be removed. In the end, the King had an abscess around the appendix drained. The operation was

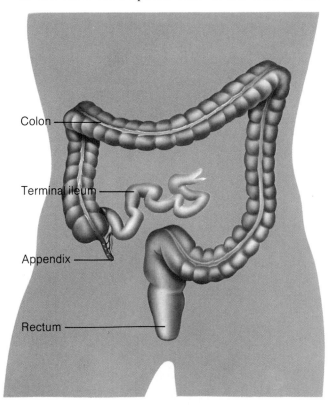

Colon

Terminal ileum

Appendix

Rectum

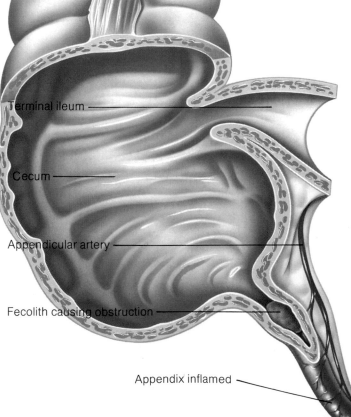

Terminal ileum

Cecum

Appendicular artery

Fecolith causing obstruction

Appendix inflamed

happened to Edward VII of England (below) *a few days before he was to be crowned, and the coronation ceremony had to be postponed while he recovered from an operation to drain an abscess around his appendix.*

performed under general anesthesia on a table in the King's dressing room, and he made a good recovery. Edward VII's case was unusual in that, although acute appendicitis can affect anyone at any age, it is commonest in children and young adults; it is rare in babies and the elderly.

The most frequent cause of appendicitis is an obstruction of the lumen of the organ. This can occur either as a result of a solid particle entering the lumen, such as a particle of feces or a fruit stone, or it can occur as a result of the swelling of the lymph tissue in the wall of the appendix. Lymph tissue often swells in reaction to infection, and this may explain why an attack of appendicitis sometimes occurs shortly after a viral infection, such as a cold or a sore throat, especially in children.

When the lumen of the appendix becomes blocked, for whatever reason, the appendix itself swells up because the secretions which normally drain out into the large bowel are unable to escape. This swelling is usually accompanied by severe pain, sometimes called appendicular colic.

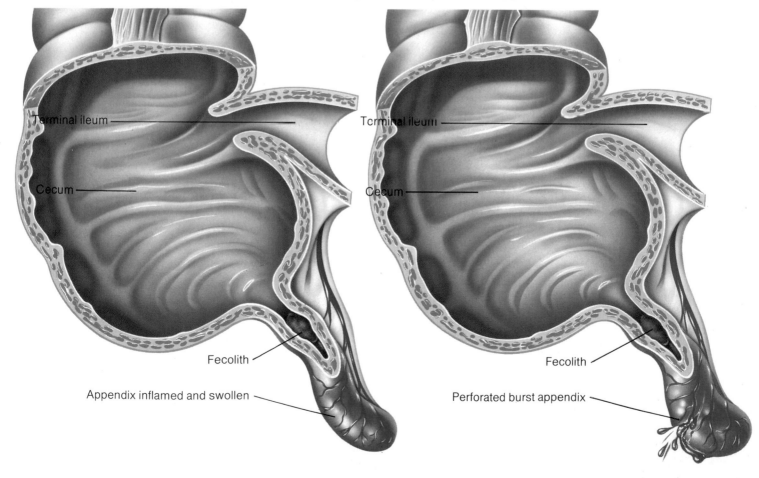

Terminal ileum

Cecum

Fecolith

Appendix inflamed and swollen

Terminal ileum

Cecum

Fecolith

Perforated burst appendix

Meanwhile, the secretions stagnate in the appendix and are infected by bacteria nearby. Eventually, the outside of the appendix becomes inflamed, and it is at this stage that a different pain, which is sharper and more localized, is felt. While appendicular colic, even though it may be severe, is a dull, diffuse pain, usually felt around the navel, this latter pain is much easier to pinpoint and is made worse by coughing or moving. It is felt in the region of the appendix, which is usually in the lower right-hand side of the abdomen—the right iliac fossa. One reason why appendicitis is sometimes difficult to diagnose is, however, because of the extreme variability in position of the appendix: it can lie almost anywhere in the abdominal cavity. Thus physicians and surgeons have to be alert to appendicitislike symptoms felt in other areas.

The structure of the colon, or large bowel, differs considerably from that of the small bowel. It is much wider in diameter at about two to two and a half inches. Also the longitudinal muscle on the outside of the colon is condensed into narrow bands, known as teniae, which when contracted throw the bowel into a series of arched folds (the word colonnade, a series of columns, comes from the same root in the Greek language). The colon is designed so that it can expand quite considerably in width and thus act as a storage place for feces. The whole colon is over three feet long and lies in the form of an inverted U so that the two limbs of the U are at the sides of the abdomen. Like the rest of the intestine, the colon is fixed to the back wall of the abdomen by mesentery.

The colon starts at the ileocecal valve — a ring of muscle at the junction between the ileum, the last part of the small intestine, and the cecum. The ileum is actually attached to the side of the colon, and the cecum is a blind pouch adjacent to this junction. The ileocecal valve is most important in preventing colonic contents from leaking back into the last part of the ileum.

The cecum is in the lower right-hand part of the abdominal cavity, and the colon leads vertically upward from here toward the liver. At the liver, the colon makes a sharp turn to the left and crosses the abdominal cavity. This part of the colon, known as the transverse colon, is attached to the posterior abdominal wall by a long mesentery so that it is relatively free to move around. In the left upper part of the abdomen, the colon makes another sharp turn and descends toward the pelvis along the left-hand side of the abdomen. It then makes an S-bend, called the sigmoid (S-shaped) colon, which ends in the rectum. This last part of the colon is about seven inches in length and terminates at the anal canal.

Bacteria in the Colon

Structure aside, the fundamental difference between the small bowel and the colon is that while the small bowel absorbs all the nutrients from the food, the only thing of significance absorbed by the colon is water. Obviously not everything that is eaten is digested, and undigested food, along with the remaining digestive juices, enter the colon, where most of the water is reabsorbed by its lining. Even when food is not being eaten, a certain amount of fluid, consisting of secretions from the lining of the small intestine and dead cells from the mucosa, enters the colon from the small intestine.

When the water has been absorbed, solid feces are formed, and these are pushed along the colon by peristalis into the rectum, where they are held until they can be evacuated at a convenient time. Because the feces remain in the colon for some considerable time, they become infected by bacteria, and a normal person plays host to many millions of bacteria in the large intestine which can be a potent source of infection of many kinds. This, and the often offensive nature of feces and flatus, is why feces are considered to be dirty. By preventing the contents of the colon from escaping back into the ileum, the ileocecal valve ensures that these bacteria do not contaminate the small bowel.

Bacterial contamination of the small bowel would lead to serious disruption of the absorption of foodstuffs, and, because the colon is full of bacteria, physicians and surgeons previously believed that it was somehow a source of pollution. Sir Arbuthnot Lane, a surgeon at Guy's Hospital in London, England, at the turn of the century, even removed the whole colon in patients with illnesses ranging from migraine to rheumatism, in the belief that it contaminated the rest of the body.

These bacteria in the colon also have another effect. Some of them produce gas, mainly methane,

116

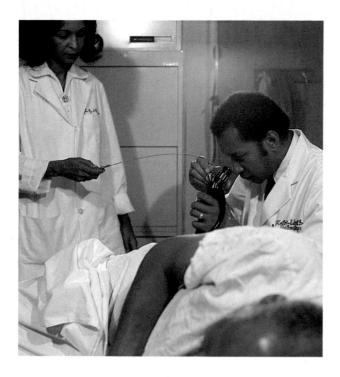

as they break down the food residue to provide energy for themselves. The variation we all notice from time to time in the production of flatus reflects — at least in part — a variation in the types of bacteria which are to be found in the colon.

However, most people also vary the types of food they eat, and this too makes a big difference to the amount of flatus produced. Beans have a justifiable reputation for the production of wind because they contain a number of trisaccharide sugars which cannot be broken down by the small intestine. These sugars reach the colon, where they encounter bacteria which are capable of breaking them down but which produce large amounts of gas as a result. In addition to the gas produced by the bacteria, there is also a certain amount of swallowed air, which travels through the alimentary system and finally arrives at the colon. The colon, therefore, not only stores feces but also stores a moderate quantity of gas. It is perfectly normal for people to release this gas at fairly regular intervals. Yet, just as people may vary greatly in the frequency with which they move their bowels, so they may also vary in the amount of flatus which they release during the course of a day.

When feces reach the rectum there may be a considerable delay before they are evacuated. In some people this delay may be several days or even weeks. It is a great tribute to the design of the large bowel and the rectum that a person can carry on eating two or three meals a day and not pass any feces for many days, without any ill effects. But modern physicians recommend that the bowel should be evacuated every few days at least.

Haustral Shuttling

Unlike the rest of the gut, the outer, longitudinal muscle layer is arranged in a straplike fashion over the colonic surface. The longitudinal muscle fibers are gathered together to form teniae. When the teniae contract, they cause the lining of the colon to fold up into sacks which are called haustra. Random contraction in the circumferential and longitudinal muscles causes alternate contraction and relaxation of these haustra, which have no effect on moving the fecal contents onward toward the anus but which are important in mixing up the contents and allowing maximum water absorption. This is a process known as haustral shuttling. Alternatively the haustra may contract down in a more or less sequential fashion that moves the feces onward.

Occasionally there is a major movement of feces by a process known as "mass movement," during which fecal matter may be transferred at a rate of over two inches per minute toward the anus. These mass movements depend upon the nerve supply to the colon which is carried in the two nerve plexuses found in the intestinal wall.

The duodenocolic reflex results from the appearance of food matter in the duodenum, which signals via the intestinal nerves to the colon, triggering a mass movement. There is also a somewhat weaker gastrocolic reflex which is triggered by the arrival of food in the stomach.

Defecation

The internal anal sphincter is normally in a contracted state so that there is no leakage of fecal material out through the anus. As the rectum becomes stimulated by the accumulation of large amounts of fecal residue, nervous impulses pass from the rectum to the anal sphincter, triggering relaxation. This sequence is known as the defecation reflex.

However, a further voluntary element is required for defecation to take place, since the external anal sphincter is under conscious control. If a person decides that he wishes to defecate, then the external anal sphincter also relaxes and defecation

Diverticulae, small saclike swellings which develop in the walls of the last part of the colon, are thought to be caused by a diet containing insufficient fiber. Diverticulitis is rare in underdeveloped countries.

takes place. It is important to emphasize that the defecation reflex is relatively weak, and that fullness of the rectum not only triggers this reflex but also triggers a voluntary desire to defecate. In everyday life the voluntary rather than the reflex consequences of rectal fullness are of greater importance.

The fact that the higher centers within the nervous system are involved in the control of defecation is borne out by the well known effect that anxiety has on the bowel habit. In susceptible individuals, the effect of anxiety is to heighten the sensitivity of the system and to lower the threshold of response to rectal filling, so facilitating the trigger to defecate. In anxiety-provoking situations, such as before tests or interviews, some unfortunate individuals may need to make many visits to the bathroom.

Diagnosis and Disease

Unlike the small intestine, the colon is subject to a number of different diseases. In some cases, of course, the colon is responding to problems which occur higher up the alimentary tract. If there is a large volume of fluid entering the colon from the small intestine, for instance when an infection makes the small intestine secrete an excessive amount of fluid, it will not necessarily be able to absorb all the water, and diarrhea will result.

Fortunately, when disease is suspected, doctors now have a number of different ways of investigating the function of the colon; many have been available for several years and some are a result of recent research. The technique of producing X-ray pictures of the colon — difficult because the colon does not show up on normal X-ray plates — was solved with the introduction of the barium enema in the 1920s. Barium — a white metallic element — inserted into the rectum and thus the colon is opaque to X rays and gives a clear outline of the colon when it is subsequently X-rayed. It passes out harmlessly in due course. This form of investigation is now standard, although it has been refined greatly since its introduction.

Barium enema examinations have been further improved by adding air so that the colon can be inflated and the lining coated with a thin layer of barium. The barium and air are introduced through

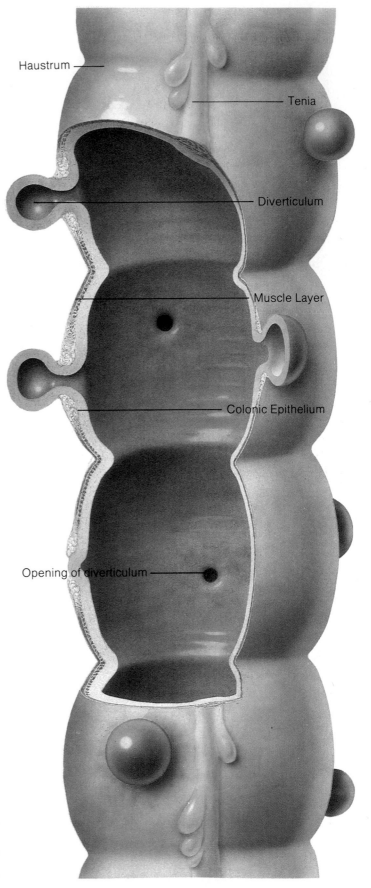

Haustrum

Tenia

Diverticulum

Muscle Layer

Colonic Epithelium

Opening of diverticulum

a tube inserted into the rectum, and the patient's position on the X-ray table is altered so that the colon is evenly coated with the barium. When an X-ray picture of the abdomen is then taken, small abnormalities of the lining of the colon can be detected. This procedure is known as a double contrast barium enema.

Diverticulitis

Nowhere is the barium enema more valuable than in the investigation of diverticular disease, one of the increasing list of diseases to which Western civilization seems to be vulnerable.

The food industry has produced steadily more refined food over the course of the last century, with the consequence that little in the way of a fiber residue is left in the colons of many people on an orthodox diet. Unfortunately this relative lack of fiber seems to increase the pressure within the colonic lumen, and this is probably the immediate cause of diverticular disease.

The Latin word *diverticulum* means a byway and it is an apt description of the small pouches which appear in later life on the side of the colon, most commonly in the sigmoid colon. Each pouch is about the size of a large pea, and there may be several dozen in a person's colon. Their exact cause is not known, but it is thought that they occur as tiny hernias of the mucosa of the colon through potentially weak points in the muscle wall where it is pierced at regular intervals by small blood vessels. Any large increase in pressure in the lumen of the colon can thus lead to a blowout of epithelium and the formation of diverticula.

Because the sacs are not lined with muscle, they become larger as the colonic muscle contracts and sometimes become filled with feces. Tiny pellets of fecal material may remain in the sac for many years until eventually the neck of the sac becomes blocked. Swelling and inflammation are the inevitable consequence, and the result is just as if a tiny appendix had become inflamed. Indeed this disease is sometimes called left-sided appendicitis, and the level of pain it produces can be severe. In the same way that an appendix can cause an abscess, so too can the inflamed diverticula, and these abscesses may rupture into the abdominal cavity, causing peritonitis. Usually, however, nothing so dramatic occurs, and sufferers have attacks of lower abdominal pain which come and go over the years.

Diverticulitis usually occurs in people over the age of fifty, but there is another disease, known by the curious name of the irritable bowel syndrome, which causes intermittent attacks of abdominal pain in younger people. The pain seems to result from the stretching of the colonic wall, which occurs as a result of higher than normal pressures within the colon. Yet again, a low amount of dietary fiber may contribute to this problem.

Compared with other parts of the world, inhabitants of the developed countries seem to have more diseases of the colon, perhaps because their diet is deficient in fiber. However, there is one colonic condition that is common in Third World countries and may even result from an excess of indigestible fiber. In some communities, especially among the Bantu of southern Africa, the sigmoid colon seems to elongate in later life. Unfortunately this can lead to a twisting of the mesentery that supports the sigmoid, and when this happens, the blood supply to this area of the gut is cut off, leading to gangrene and death if the gangrenous sigmoid bowel is not removed by surgery.

Colonic Cancer

Another way in which the large intestine differs from the small intestine, and one of the most important in medicine, is in its predisposition to the development of cancer. Cancer in the small intestine is extremely rare, but cancer of the colon is common; in fact, it is the fourth most common site of cancer, after the breast, lung and stomach.

Fortunately cancer of the colon responds well to surgical treatment if it is picked up in its early stages. The first sign is usually the appearance of blood in the stool. Even if there is no obvious blood present, there will be detectable amounts found on laboratory screening. The presence of "hidden" or occult blood is the basis for screening tests for cancer of the colon, and it may soon be routine to screen whole populations, as smear tests are now widely done to detect cancer of the cervix in women. Another symptom which should be investigated is a change in an individual's pattern of bowel evacuation.

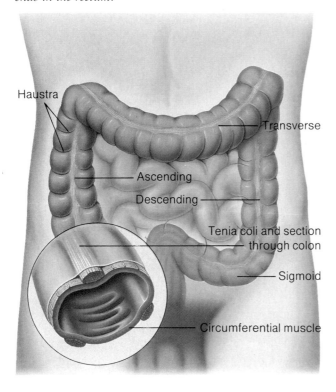

Haustra

Transverse

Ascending

Descending

Tenia coli and section through colon

Sigmoid

Circumferential muscle

If there is any clue that cancer of the colon is present, either as a result of screening or because of the development of symptoms, investigation is fairly easy. Obviously the barium enema is useful to physicians, but the expansion of fiberoptic technology has meant that, theoretically at least, it is just as easy to look into a colon as it is to look into a stomach or a duodenum. The instrument now used, the fiberoptic colonoscope, is about five feet long, longer than the gastroscope. While it is more difficult to steer a fiberoptic instrument around the bends in the colon than it is in the upper gut, excellent views of the colonic mucosa can nevertheless be obtained. The colonoscope can also be used to take samples of areas of abnormal mucosa, and polyps that may be seen growing out of the mucosa can be removed by snares extended from the colonoscope. The presence of a polyp can have great significance, since it is thought that most colonic cancers start off in this way.

Polyps

Small lumps, or polyps, are sometimes found attached by a stalk to the lining membrane of the colon. Most polyps are harmless in themselves, but some have the capacity to grow and turn into malignant tumors; indeed, many doctors feel that any polyp greater than half an inch in diameter is likely eventually to turn into a cancer. Polyps can be solitary, or they can be present in great numbers. There is a condition known as *familial polyposis coli* in which hundreds of polyps develop in the colon, usually during adolescence. The condition is hereditary, and it is virtually inevitable that one of these polyps will develop into cancer later in life. Any patient who is found to have this condition should be closely questioned about similar problems in members of his family, and all close relatives should be sought out and investigated to see if they, too, are sufferers.

Single polyps are usually removed with a wire snare which is passed up through a colonoscope, but in the condition of *familial polyposis coli*, the whole, or nearly the whole, of the colon must be removed to obviate the likelihood of one of the polyps turning cancerous.

Many theories on the development of cancer of the colon have been advanced but none has been proved beyond doubt. One school of thought is that the time the feces spend in the colon is important. A long transit time increases the length of time during which the feces are in contact with the lining membrane and may increase the risk of cancer. This theory fits in with the fact that cancer of the colon is uncommon in countries where a high fiber diet is normal. This type of diet is known to speed up the transit time. Other theories involve the content of the feces, and there is some statistical evidence to suggest that eating a great deal of meat is associated with a higher incidence of cancer of the colon.

Whatever the cause, operations for cancer of the colon or rectum vary only slightly, depending on the exact site of the tumor. Essentially they involve the surgical removal of the tumor together with a margin of normal colon or rectum on either side of it. The two ends of the colon or rectum are then stitched together to restore the continuity of the bowel. Because this juncture will have bacteria-laden feces passing through it, any leakage at the site could have extremely serious consequences in terms of infection and peritonitis. Therefore, in some cases, where there is doubt about the integrity of the join, a temporary colostomy is performed.

A colostomy is a device whereby part or all of the colon is brought through a hole in the front of the abdomen and stitched to the skin. The feces then pass out into a bag, which is attached to the skin with special adhesive. The colostomy is made upstream from the join in the colon, so that the feces escape through the colostomy and do not pass through the area where the bowel has been joined up. Only when the tumor is situated in the lowest part of the rectum so that the last part of the rectum and the anal canal have to be removed does the patient need to have a permanent colostomy. A new technique, whereby the ends of the bowel are joined together with metal staples, has now made it possible for some of these low rectal cancers to be removed and the ends of the bowel connected, obviating the need for a colostomy.

With the advent of plastics and new adhesives, a colostomy is no longer the horror that it was only thirty years ago. Then, a patient with a colostomy had a constant problem with leakage and odor, and many patients found it virtually impossible to carry on a normal social and sexual life. Nowadays, a perfect seal can be made between the bag and the skin so that there is no leakage of gas or feces.

It was a Frenchman, Desormeaux, who first started looking at the rectal mucosa in 1853, using a primitive sigmoidoscope that relied on an alcohol and kerosene impregnated wick as a light source. Over the next fifty years, the technique of sigmoidoscopy was perfected. In those days tuberculosis was rife, and as a result of this infection the bowel mucosa frequently presented an inflamed surface appearance; there might even be obvious bleeding. As time went on and tuberculosis became less common, there were still many patients who had an obviously inflamed lining to their bowel. But despite diligent searching, physicians found no evidence that the inflammation resulted from infection, and over the years, they have come to refer to this condition as inflammatory bowel disease.

Crohn's Disease

There are two major diseases involved in inflammatory bowel disease, ulcerative colitis and Crohn's disease, which was first described by

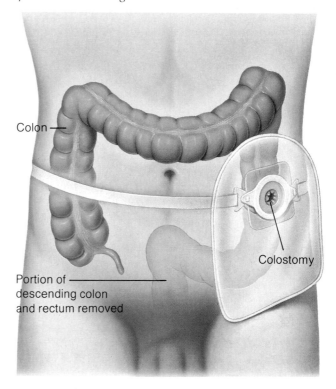

With a colostomy, feces pass through a hole in the front of the abdomen and into a bag. A perfect seal is necessary between the skin and the colostomy bag (below) to prevent problems with leakage or odor.

Colon

Colostomy

Portion of descending colon and rectum removed

Burrill B. Crohn in the 1930s. It is a great tribute to Dr. Crohn, who died in 1983 at the age of ninety-nine, that he recognized this disease as a separate entity to tuberculosis, since tuberculosis was so common at that time. Nowadays, Crohn's disease stands out very obviously, for there is virtually no incidence of intestinal tuberculosis to confuse the picture.

Crohn's disease can, in fact, affect any part of the alimentary tract from the mouth to the anal canal, but it is commonest in the last part of the small intestine, known as the terminal ileum. There may be only one patch of the disease, or there may be several separate patches in different parts of the large and small intestines. These patches are known as skip areas and may vary in severity.

The fundamental feature of Crohn's disease is an inflammation of the full thickness of the bowel wall, from the inner lining right through to the outside. The inflammation causes swelling of the bowel wall and ulceration of the lining, with the result that parts of the lining are missing and bleeding can occur from these raw areas. As the disease progresses farther, two things occur. First, the swelling

Burrill B. Crohn

An Interest in Indigestion

When Dr. Burrill B. Crohn was asked by an interviewer why he had chosen medicine as a career, he replied: "Originally, because my father had terrible indigestion. I decided to help him by studying medicine." This decision led to Crohn becoming known worldwide for his description of the condition known as Crohn's disease.

Crohn presented his description of fourteen cases of regional ileitis — the condition of inflammation of the bowel that now bears his name — to the Section of Gastroenterology and Proctology of the 83rd Annual Meeting of the American Medical Association on May 13, 1932. A number of experienced clinicians in the audience commented on similar cases they had seen, and, indeed, probably the first description of this condition was given by Giovanni Battista Morgagni (1682–1771), who was a Professor of Anatomy in Padua for over fifty years. Nevertheless, there can be no doubt that the essential work leading to the recognition of this disease was the paper written by Crohn, Ginzburg and Oppenheimer in 1932, and that most previous cases had probably been misdiagnosed as tuberculosis.

Crohn believed that the condition was restricted to the

terminal ileum, but the description regional ileitis was preferred to terminal ileitis to avoid patients' worries that the word "terminal" referred to their prospects, not merely to the distribution of the inflammation. It was another twenty years after Crohn's description before it was recognized that the terminal ileum was not the only part of the gastrointestinal tract to be affected by this condition.

There have been countless publications on the subject of Crohn's disease since that first description, and Crohn subsequently published a book on the disease, reviewing 1,100 of his patients. However, we have not got any closer to discovering the cause.

Born on June 13, 1884, in New York City, Crohn received a B.A. degree from the City College of New York and, in 1907, his medical degree from the Columbia University College of Physicians and Surgeons. His professional life was spent at the Mount Sinai Hospital, New York. He was a highly respected gastroenterologist, and during his career was awarded the Townsend Harris Medal by City College in 1948, the Julius Friendenthal Medal of the American Gastroenterological Association in 1953 and the Mount Sinai Hospital's Jacobi Medal in 1962.

Crohn published widely — his 1927 monograph on *Afflictions of the Stomach* was particularly highly regarded. He was President of the American Gastroenterological Association in 1932, and was said to be ambitious and sometimes headstrong. However, he was also a modest man and at a conference in Prague he opposed a resolution officially designating ileitis as Crohn's disease. His objection was ruled out and the resolution was otherwise adopted unanimously.

After his retirement, Crohn lived in New York, pursuing his special interest in the history of the American Civil War, until his death at the remarkable age of ninety-nine, on July 29, 1983.

In Crohn's disease, ulceration of the lining of the bowel wall results in the lining being eroded (below). If left untreated, the ulcers can penetrate right through the wall of the bowel into an adjacent organ.

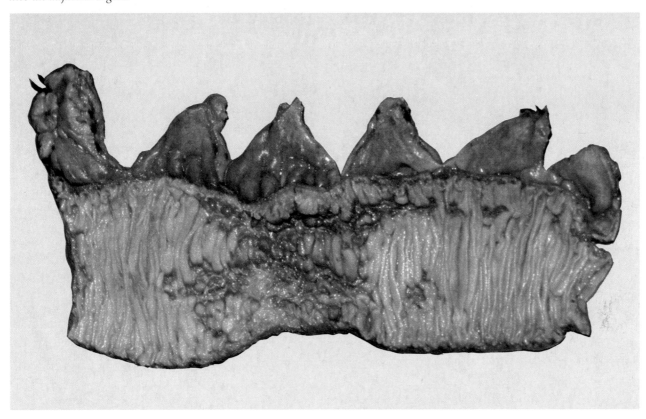

of the bowel wall leads to narrowing and rigidity so that eventually a blockage occurs. But before there is a complete blockage, the narrowing causes a severe delay in the passage of intestinal contents, with consequent pain and bloating, usually within an hour of eating a meal.

Second, the ulcers of Crohn's disease may eventually penetrate right through the wall of the bowel and into either an adjacent organ or into the wall of the abdomen. If the ulcer penetrates an adjacent organ, an abnormal link may be formed between two loops of the bowel, or between the bowel and, say, the bladder. If the ulcer penetrates into the wall of the abdominal cavity, the patient may develop a passage between the bowel and the outside of the body. One day, a swelling may be noticed that eventually breaks down like an abscess, but instead of pus being produced, intestinal contents begin to discharge onto the surface.

Crohn's disease can vary greatly between individuals, and though it affects mainly young adults, it can also occur in children and old people.

Its course and outlook depend on how severe the inflammation is and how many separate areas of the intestine are affected. In a patient with one area of Crohn's disease in the ileum, there is about a forty percent chance of recurrence of the disease if that area is removed surgically. Thus, there are many patients who have only one episode of the disease in their lifetime, and there is another group of patients who have repeated attacks. The cause of Crohn's disease is unknown. Many theories have been put forward, but so far no one has discovered any definite cause.

Treatment

The treatment of Crohn's disease is either by drugs or by surgery. Close attention is also paid to nutrition, which may be severely impaired by the disease. Drug treatment includes the use of cortisonelike drugs, which have the capacity greatly to modify the body's inflammatory process, and Azathioprine, which reduces the immune response of the body. Both can cause a decrease in the

An internal hemorrhoid is one that occurs near the beginning of the anal canal. Hemorrhoids are rich in blood and bleed easily. They are thought to result from a diet which contains insufficient fiber.

External hemorrhoids (below) develop at the edge of the anal canal. These can swell and cause the sufferer considerable pain during defecation, and fear of defecation can, in turn, lead to constipation.

In the condition known as anal fissure (below), the skin of the lining of the anal canal becomes split during defecation, causing bleeding and severe pain. Often, anal fissures heal spontaneously.

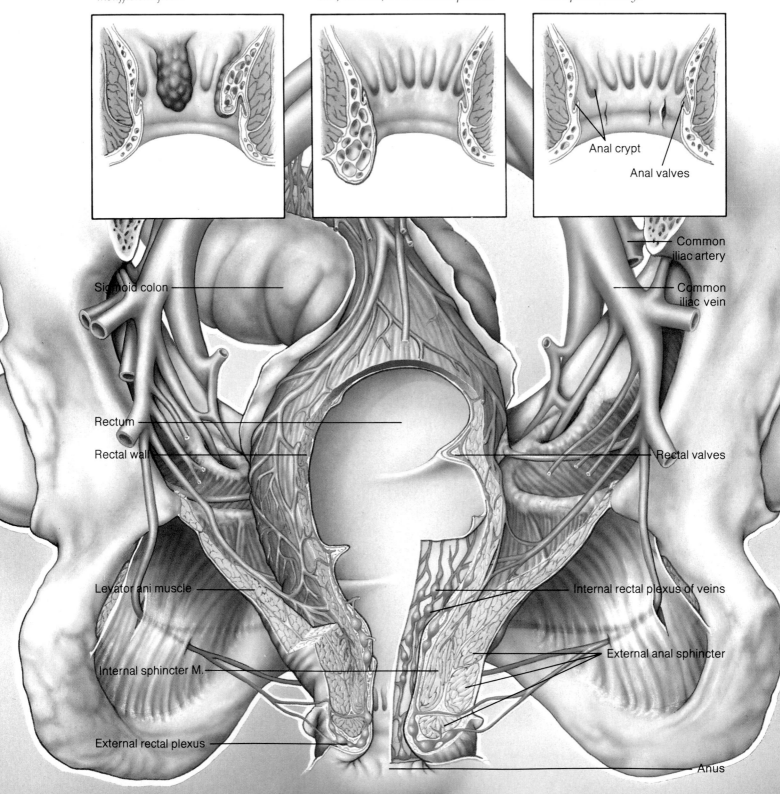

Anal crypt

Anal valves

Common iliac artery

Common iliac vein

Sigmoid colon

Rectum

Rectal wall

Rectal valves

Levator ani muscle

Internal rectal plexus of veins

Internal sphincter M.

External anal sphincter

External rectal plexus

Anus

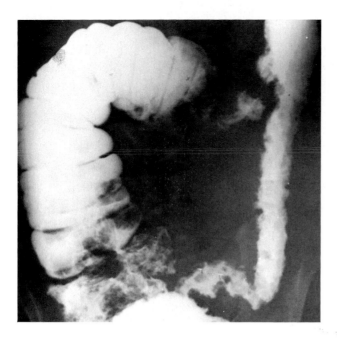

The ragged appearance of the descending colon seen in the X ray (left) indicates a condition known as ulcerative colitis. This can affect the entire large intestine and requires prolonged treatment.

'amount of inflammation, leading to a significant improvement in a patient's symptoms, but it is doubtful whether either of these drugs can lead to the disappearance of a segment of Crohn's disease. They are, however, extremely useful, especially in patients with multiple areas of the disease.

Surgery is employed in Crohn's disease when there is severe narrowing of the bowel or when a passage exists between the affected segment and another organ. The inflamed portion of bowel is removed, and the remaining bowel is joined together again.

Ulcerative Colitis

As its name implies, this disease affects the colon and consists of an inflammation of the lining membrane only; penetration of the bowel wall by ulcers, which occurs in Crohn's disease, does not take place. It can affect a short segment of colon, or the whole colon, but it always affects the last part of the large bowel — the rectum. In other words, a patient may have ulcerative colitis affecting the whole colon, but a patient never has ulcerative colitis affecting just the ascending part of the colon, with a normal rectum.

Although the disease affects young adults most commonly, it can occur at any age and it is slightly commoner in women. It is less common in Negroes than in Caucasians and is uncommon in Japan. Many factors have been thought to cause it, including diet, stress and allergy, but none of these connections has been proven. There appears to be a slight tendency for it to run in families, but it is not strictly an hereditary disease.

The symptoms of ulcerative colitis are diarrhea, and the passing of blood and mucus. Other symptoms include abdominal pain, which usually precedes the diarrhea, and a constant feeling of incomplete evacuation. If the disease is severe and of long standing, then the symptoms of anemia (tiredness, lassitude and shortness of breath) may be present. Symptoms may come and go without any treatment, and even severe symptoms may suddenly disappear only to return a year or so later for no apparent reason. It used to be thought that stress caused this disease, but whereas that is no longer an accepted theory, there is no doubt that stress can trigger off attacks in some people or can make an attack worse. Curiously, there is now a little evidence to suggest that cigarette smoking may protect people against the disease.

Most cases of ulcerative colitis respond well to treatment with drugs, either taken by mouth or as liquid preparations administered via the rectum. The two main drugs used are Azulfidine, which is related to aspirin, and Prednisolone, which is related to cortisone. The way in which the colon responds to treatment can be assessed by looking at the mucosa with a sigmoidoscope or colonoscope.

If the disease fails to respond to medical treatment, or if any of the complications of the disease occur, then surgery may be needed. Surgery usually involves the removal of the whole or part of the large bowel and may necessitate the formation of an ileostomy or colostomy. Although this may seem a drastic measure, some patients' lives are transformed dramatically following surgery. They may have been constantly bothered by diarrhea and abdominal pain, which disappear when the affected segment of bowel is removed.

Finally, it should be said that Crohn's disease and ulcerative colitis are related diseases, both involving inflammation of the bowel, for which no cause can be found. In the early stages of their disease, it is impossible to place some patients in one category or the other, and their complaint must merely be labeled by the physician as nonspecific inflammatory bowel disease.

Anal Canal

The difficulties that patients with poor colostomies used to suffer highlight the importance of the

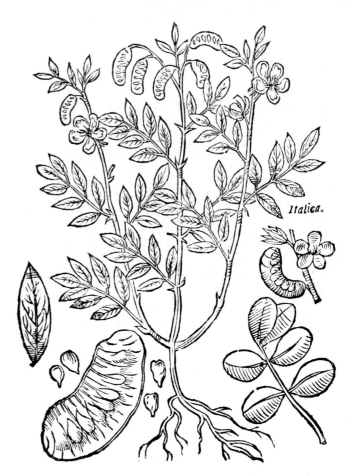

Senna pods and dried senna leaves (right) have long been used as a cathartic and laxative. Many people mistakenly believe that to remain healthy they must have a bowel movement once or more times a day.

control of defecation, which is the province of the anus. The anal canal is formed by a ring of muscle around the junction between the rectum and the skin, and if the canal is not functioning normally, severe psychological and social problems may ensue. It has already been mentioned that the function of the colon is to store up waste products until the time comes to eliminate them, and the anal canal is the final, sophisticated switch that in normal circumstances prevents the elimination of flatus or feces until the time is convenient.

During the development of the intestine in the embryo, the end of the intestine grows down into the pelvis and joins up with a dimple of skin between the legs. Eventually, all that separates the rectum and the skin is a thin membrane, or diaphragm, that eventually breaks down. Thus the anal canal itself is lined partly by skin containing nerves sensitive to painful stimuli. The whole canal, which is about two inches long, is surrounded by a tubelike ring of muscle, which is partly under voluntary control. During defecation, the muscle lining the rectum contracts, while the muscle of the anal canal relaxes, allowing feces to be passed. There is also a clever sensory mechanism — not fully understood — which distinguishes feces from flatus and allows only flatus to be passed at certain times.

The anus is subject to a number of diseases, many of them fairly trivial, but most either painful or embarrassing to the sufferer. Fortunately for physicians the anal canal is easily visualized, using an instrument called a proctoscope. This is little more than a piece of metal tubing, only a few inches long, through which a light source can be directed. Like many a simple tool available in a physican's office, it is immensely valuable. The various conditions of the anus are relatively simple to diagnose because the area is so easily visible through the proctoscope, and surgeons are also able to carry out simple operations through it.

Physicians have also been looking at the mucosa beyond the upper end of the anus for over a century, not by using the modern fiberoptic colonoscope but by using a simpler, rigid tube about a foot long, which is no more than a longer version of a proctoscope. This is called the sigmoidoscope, but its name is misleading, since

it is very difficult to maneuver the instrument right up into the sigmoid colon. The instrument is generally used to observe the first six to eight inches of rectal mucosa.

Disorders of the Anal Canal

There are several disorders of the anal canal which, because of social inhibitions, tend to become mixed together in people's minds. In fact, almost every disorder of the anal canal has at some time been referred to by the sufferer as piles. This has led to many months of inappropriate, and sometimes actually dangerous, treatment.

Although many of the minor conditions of the anal canal can cause hours of misery, they are not in themselves dangerous. The most important pitfall for physicians is that some of these can mimic cancer, leading to delay in diagnosis.

Cancer of the anal canal is a relatively uncommon but serious disease. When it does occur, the tumor usually arises from the skin part of the lining of the anal canal and is rather like a skin cancer. However, it is considerably more serious than straightforward skin cancers which can have a cure rate of more than ninety percent. Because of delay in diagnosis (the patient attributing any bleeding to piles) and because the area is rich in lymphatics, the tumor

126

La Chasse (below) *by Carle van Loo depicts the lavishness of meals eaten by the rich in Europe in the eighteenth century. In the 1980s, people are becoming aware of the problems caused by overeating.*

may spread, and radical surgery is usually needed to effect a cure. Even then, the proportion of patients surviving five years or more is only in the region of forty percent.

Some children are born with abnormalities of the anal canal and rectum. The rectum may have failed to grow down into the pelvis, or the diaphragm between the skin and the rectum may fail to break down, leaving a complete blockage and rendering the child incapable of passing flatus or feces. In others, the anal canal may open in an abnormal site, usually more to the front, so that it may open into the vagina in a girl or into the urethra in a boy. Most of these abnormalities can be dealt with surgically, although they may require several operations and the formation of a colostomy before the treatment is complete.

Hemorrhoids

Fleshy tags, or hemorrhoids, arise from the upper part of the anal canal and, as they enlarge, may protrude out of the anal canal during defecation. There are usually three of them and their cause is a mystery, although there is no doubt that they are

yet another disease of Western civilization, since again they are uncommon in areas where the diet contains considerable roughage. These internal hemorrhoids are not the same as the external tags of skin, or external piles, which some people develop at the edge of the anal canal.

Because internal hemorrhoids are rich in blood vessels, the main symptom in the early stages is bleeding; later on, when they enlarge, they may be obvious as lumps which are felt by a sufferer after defecation. Occasionally the lumps swell up and become extremely painful. This may be called an attack of piles by the patient, but the physician will call it thrombosed hemorrhoids. Usually, the pain abates after a few days of treatment with warm baths and icepacks, but occasionally surgery is necessary if the symptoms fail to disappear.

Most cases of hemorrhoids are still treated by conservative or nonsurgical methods. The commonest form of treatment for small hemorrhoids, which only bleed, is to inject them with a liquid that causes them to shrink. This is a painless procedure and can be performed easily in the physician's office. Other treatments that can also be undertaken in the office include freezing by the application of a cryoprobe, and the application of a tiny rubber band around each hemorrhoid. This band cuts off the blood supply to the hemorrhoid, hopefully causing it to shrivel up. If these methods fail, then surgery can be undertaken, but it is really only needed if the hemorrhoids are very large.

Anal Fissures

A condition which is often confused with hermorrhoids and, therefore, treated wrongly, is anal fissure. In this condition, the skin part of the lining of the anal canal becomes split or torn during defecation. Because this skin is very sensitive, there is severe pain, and there may be a slight amount of bleeding. The pain occurs only during defecation but may be so severe that the patient is afraid to move his or her bowels. This then leads to constipation, which makes matters worse.

Sometimes anal fissures heal spontaneously with the aid of some local anesthetic cream, but in severe cases minor surgery is required. This takes the form of a small cut in the lowermost fibers of the anal canal muscle, which relieves the intense muscle spasm in the region of the fissure. Relief of the spasm allows the fissure to heal up and allows the patient to defecate with less effort.

Abscesses and Fistulae

Another condition of the anal canal which sometimes causes concern is when an abscess — a cavity containing pus — forms as a response by the body to invasion by certain types of bacteria. In the anal canal, there are several small glands that provide lubrication to the passage of feces and sometimes the opening of one of these glands becomes blocked, leading to swelling and the formation of an abscess. This will cause pain for several days and will eventually burst, possibly into the anal canal itself. But, more troublesome, it may also burst through the skin next to the anal canal, leading to the formation of a fistula, or abnormal track, connecting the anal canal with the skin. As can be imagined, this leads to troublesome discharge, which may persist for months or even years. At one time, fistulae were extremely common, but nowadays they are far less frequently found because of improved surgical treatment.

Chapter 7

Maintaining the System

In times gone by, infectious disease was the most feared of all the afflictions of mankind. The spread of disease from the stricken to the healthy seemed almost supernatural, and myth and magic surrounded this terror of infection. During the fearful cycle of epidemics which ravaged Europe during the fourteenth century, around twenty-five million people died — a quarter of the continent's population. Plague was the most feared killer — villages and cities could be wiped out by the arrival of a single sickly traveler. Many European epidemics came from the Far East, often through the southern French port of Marseilles. In later centuries, smallpox took over as the most devastating infection. More recently, in the early years of the nineteenth century, when peasants were moving into industrialized cities in search of work, cholera and typhoid added their terrors to community life.

Mankind's long struggle against disease has been a tale of chance discovery, painstaking research, inspired deduction and, on the negative side, occasionally inexplicable adherence to outmoded beliefs and premises. One of the most fortunate events occurred in 1673, when a Dutch draper, Anton van Leeuwenhoek, built a simple microscope and so was able to observe and describe the movements of bacteria and simple one-celled organisms called protozoa. His discoveries — ahead of their time — were not, however, linked in any way with diseases and their transmission. In fact, it was not until two centuries later that Louis Pasteur realized the true association between certain microbes and infections.

Certain principles were observed in the treatment of infections. One of these was the need to isolate the sick. This was realized many centuries ago, and strangers traveling from distant lands were sometimes quarantined. In 1666, the inhabitants of a small village called Eyam, in the Peak District of Derbyshire, England, isolated them-

Cholera, allegorized here as Death the Reaper, *not the Bulgarian army was the main enemy of the Turkish army in 1912. As the Turks advanced, cholera was taking up to one hundred lives each day.*

Monster soup is an apt name for this eighteenth-century satirical impression of the polluted river Thames in London (below), for these waters were full of the bacteria which cause cholera.

In many countries restrictions are in force to prevent industrial waste or untreated sewage from being discharged into rivers or oceans, and samples of water are constantly checked for pollution (bottom).

Pasteurization, a process invented by Louis Pasteur, a French chemist, (right) destroys harmful micro-organisms in food. This process is still widely used today in commercial food processing.

MONSTER SOUP commonly called THAMES WATER, being a correct representation of that precious stuff doled out to us !!!

selves from the rest of the country when a resident was struck down with plague — two-thirds of the village died, but the surrounding areas remained free from infection.

Another aspect of disease control is exemplified by the work of Edward Jenner, an English country doctor and naturalist. At the end of the eighteenth century, he observed that milkmaids who had suffered from cowpox — a comparatively mild illness — had an immunity to the killer smallpox.

He took some of the lymph from cowpox pustules and scratched the lymph matter into the skin of a young boy. When he inoculated the boy a few weeks later with matter from a smallpox pustule, he did not develop even a mild case of smallpox. American President Thomas Jefferson wrote to Jenner: "Future nations will know by history only that the loathsome smallpox has existed and by you it has been extirpated."

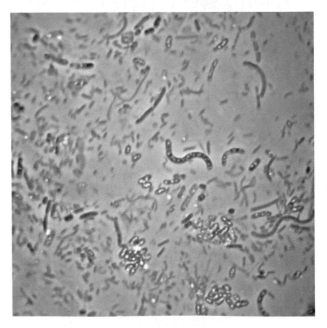

Nature, however, has a way of exploiting what are apparently advances in the lot of humankind and springing nasty surprises. For instance, in the nineteenth century, maritime trade routes were opened between Europe and North America, the Old and New Worlds, and this gave excellent opportunity for the transmission of deadly infectious diseases. The construction of the Panama Canal was almost stopped by a mysterious and lethal disease, now known to be yellow fever, which spread from South America as far north as Philadelphia in 1790 and was responsible for ten thousand deaths in that city.

The control and conquest of diseases such as smallpox and cholera has been difficult, and of primary importance has been the introduction of public health measures, improved sanitation and the cleansing of the environment. Increased knowledge of the infecting agent and the development of vaccines have also made enormous contributions. But these measures have been a comparatively recent development; in the U.S.A. there were no real public health laws, apart from a few smallpox regulations, until 1849. In that year, after a second huge epidemic of cholera in Massachusetts, a Sanitary Commission, led by Lemuel Shattuck (1793–1859), was set up. He stressed the large amount of ill health caused by insanitary conditions, but it was not until 1869 that the Massachusetts State Board of Health was established, the first permanent organization of its kind.

Cholera

At about the same time, the London physician John Snow showed by an ingenious epidemiological survey that cholera was spread by water and not by mists and fogs as was previously thought. By

marking on a map the location of each case of cholera in his practice, he was able to show, in 1854, that the pattern of cholera cases corresponded to the area of distribution of drinking water from just one hand pump in the city. He removed the handle from the pump, and the disease diminished. Noting the contaminated state of the water from the pump, he suggested that sewage and drinking water must at all times be separated; this principle was not fully appreciated before that time.

Old beliefs die hard, however, and in Malta, a few years before this, the government had declared that cholera was not an infectious disease. Indeed the theory of contagiousness in cholera was said to be a "cruel and unfounded" doctrine. Believers in such a doctrine were even disqualified from holding public office. Bloodletting was used to "let out" the cholera, and leeches were often applied to the abdomen or around the anus. The disease continued, and no scourge was feared more than cholera in the early nineteenth century.

Even in the 1980s, in spite of vaccinations and a huge understanding of the epidemiology of cholera, there are still countries which have been unable to eliminate the disease. So cholera remains a dangerous foe, and a close look at it will show its deadly effects.

Cholera is caused by a rapidly moving bacterium, the vibrio, which is shaped like a comma. In the course of the illness a patient may excrete between one and three-quarter pints and twenty-four gallons of watery diarrhea, containing, in the acute stages of the disease, over a million vibrios in each milliliter of stool. Unless there is strict hygiene, the disease passes readily from one person to another in contaminated water or food supplies. Some people excrete the vibrios but exhibit no disease symptoms. An example of food contamination occurred in Jerusalem in August 1970 and probably in many other places: sewage polluted water had been used to spray vegetables when they arrived at market slightly wilted from the journey. The

bacteria were able to survive only a few days on the fruit and vegetables, but were a definite health hazard. In milk and butter, the vibrios survive for up to a month or even longer. Fish and shellfish, if eaten raw or not properly cooked, have been the source of many an outbreak of cholera in the Mediterranean, when the water in which they were caught has been infected by human excreta.

When this deadly disease has gotten a hold on a victim, treatment depends upon replacing, either by mouth or intravenously, the fluid which has been lost. Salt (sodium chloride), bicarbonate and potassium also need replacing, and glucose has been found to reduce the volume of diarrhea. Antibiotics may help to control the spread of the disease but do not really hasten recovery. Cholera vaccines so far in use are likely to be effective in only fifty percent of clinical cases, and vaccinated people, although free from symptoms, may still excrete the cholera vibrios. By the mid 1980s the most effective measures for controlling the disease remained ready treatment for patients and sanitary control of the environment.

Food Poisoning

As the name suggests, food poisoning is an illness caused by food contaminated with poisons, or hostile bacteria, or when the food itself is poisonous, as in the case of the toadstool. Also, some foods are normally perfectly digestible but in certain circumstances become poisonous. Potatoes, for instance, normally contain a harmless amount of the alkaloid poison solanine. However, if the potatoes have sprouted or have been exposed to light for a long period, the concentration in the skin and just under it may rise to a level where a toxic effect becomes possible. The alkaloid, which makes the potato green is soluble in water so that if potatoes are peeled and then boiled they will contain little solanine when eaten. But if the potatoes are baked in their skins and then eaten, symptoms of solanine poisoning may develop within a few hours. Usually these consist of headache, fever and abdominal pain, often with severe vomiting and diarrhea. Recovery occurs within a few days, but the potato has claimed a few victims.

Bacteria and the salmonella group of microbes are frequent contaminants of food. Particularly feared

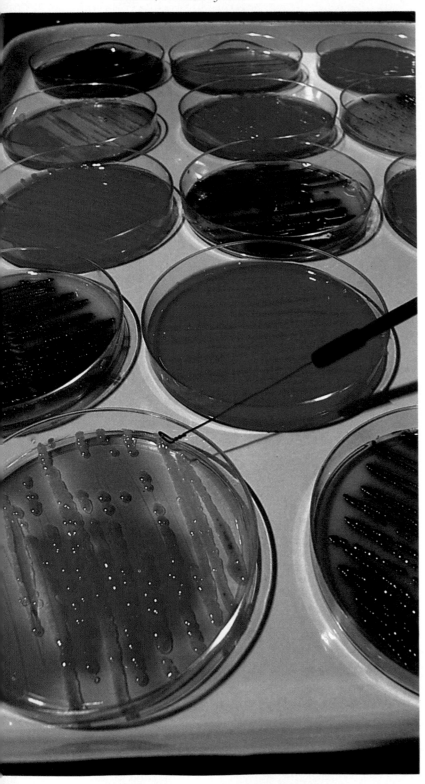

Salmonella typhosa, *shown here in laboratory cultures, is found in contaminated food that is uncooked, undercooked, reheated or cooked when only partially defrosted. It can, however, be killed by heat.*

is typhoid fever, caused by a virulent bacterium of the salmonella type. Salmonellae are most frequently found infecting chickens and their eggs, and cow's milk. Fortunately they are readily killed by heat — twenty minutes at 140°F — so pasteurization and baking kill all salmonellae in eggs and milk. But if utensils used in the preparation of uncooked food have been contaminated and are then used for beating up pasteurized cream, for example, the cream will become contaminated. Moreover, it is essential that heat must penetrate right into the center of the food to cook it fully: frozen chickens must be thawed completely before cooking, otherwise the danger from uncooked, infected flesh still remains at the end of the cooking time.

In a slaughterhouse, meat can be infected either by contamination after the animal has been killed or when live animals are infected by others in the holding pens. Frequent checks are carried out and slaughterhouses must be extremely clean and well designed to minimize the risk of cross infection and contamination.

Even seemingly innocent foodstuffs can store salmonellae. In 1973–74 in the United States and Canada, more than two hundred cases of food poisoning caused by salmonellae were traced to a chocolate factory. The source was the raw cocoa beans in the factory; the bacteria contained in them were not destroyed by the heat used during the process of preparing the chocolate, and the salmonellae survived for many months. But it is not only raw materials that cause salmonellae to spread; it has been shown that they survive in the feces of rats for 150 days, and these unpleasant creatures can infect foodstuffs in factories and stores. Cockroaches can, similarly, be agents in contaminating foods, as can flies, which have such an easy access to both bakeries and bathrooms.

Another bacterium causing food poisoning, the clostridium, produces spores which are resistant to temperatures over 212°F. If contaminated meat is allowed to cool slowly, the spores of this bacteria rapidly germinate and the bacteria multiply extremely quickly, producing a toxin which induces vomiting and diarrhea if the meat is eaten. If meat or stew is not to be eaten immediately after cooking, it should be cooled quickly, then refrigerated.

Yet another contaminating agent are the staphylococci, bacterial parasites of animals and man. Types that cause food poisoning are found on the hands and noses of between twenty and fifty percent of normal people, so towels, tables, knives and dishes are easily contaminated. Flies also are often infected and carry contamination to food. Milk is frequently contaminated in the cowshed, but the pasteurization process destroys the bacteria. As they multiply, the bacteria produce a toxin which will withstand boiling for thirty minutes, so even if the bacteria are killed, the food remains dangerous for some time.

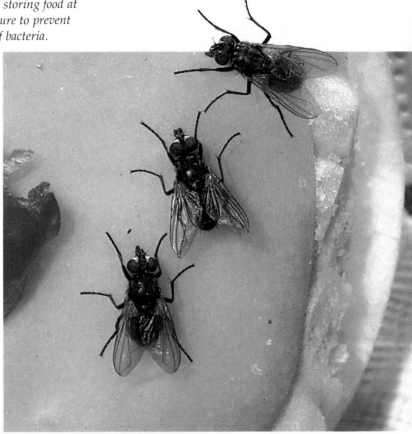

With all these potential hazards around, two important rules must be observed in the food industry. First, there must be no contact between cooked and uncooked food, whether poultry, vegetable or dairy products. Second, if bacteria do come into contact with food at some stage during its preparation, they should be given no opportunity to multiply. The temperature of the food is the key: it should be either too hot or too cold for bacterial multiplication. As well as applying these two principles, staff must be intelligent and watchful, and scrupulous cleanliness is, of course, essential.

To sum up, while humans are made up of a countless collection of cells, and each potentially fatal bacterium is comprised of only one, until recently we have been at the mercy of malicious microbes. The history of epidemics is the history of wars, traveling, poverty, famine and drought. Improvements in sanitation and hygiene have been the most effective measures in stopping the spread of disease in the Western world during the course of the last century.

A History of Dentistry

Infectious disease, especially that carried by food and water, has perhaps been the most important cause of death in bygone years, but the teeth held pride of place as the cause of everyday discomfort. Few people are immune from the ravages of toothache, and knowledge accumulated over the last hundred years has had its emphasis on prevention. Few practitioners can be filled with as much justifiable pride in the incisive manner in which they have crossed the bridge from repair to prevention as can dentists.

In Elizabethan England, at the time of Shakespeare, the rich suffered agonies from toothache, due to an excess of sugary cakes and sweetmeats and lack of oral hygiene. Queen Elizabeth herself had a mouth full of blackened stumps. It is known that she had some of them extracted, and subsequently, for appearance's sake, used a device to improve her appearance which can have been effective only when she did not open her mouth. As was observed by Herbert Norris in 1602: "The Queen is still . . . frolicy and merry, only her face showeth some decay, which to conceal when she cometh in public she putteth many fine cloths into her mouth to bear out her cheeks."

Nearly two centuries later, George Washington put rolls of cotton into his mouth when posing for the portrait still seen on today's paper money and stamps. There is a certain fullness around the chin and lower cheeks where there should have been a concave curve, and he has his lips tightly closed. This padding was used to soften the contours of an unattractive set of false teeth. Dentistry has been

138

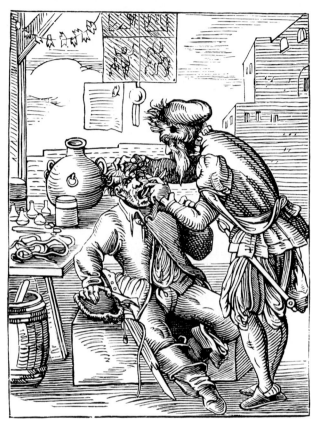

Dentistry in the Ancient World

Many thousands of years ago, human beings had to chew their meat raw, gnaw at bones, and pick out stones from among the vegetables and nuts grubbed from the earth, and even in these primitive times, tooth decay was not uncommon. In a study of thirty-two skulls dating from the Bronze Age, seven were found to have tooth decay. The grit and shells that came with the ancient diet simply wore away the surface of the teeth so that bacteria could enter the soft pulp and cause abscesses.

Five thousand years before Christ, the Babylonians believed that toothache was a punishment by the gods, and that relief could be brought about only by prayers and incantations. The ancient Greeks — more practical and understanding — made pliers for extracting teeth and also tied loose teeth to their adjacent sound neighbors with gold wire. The most advanced, imaginative dentistry of the ancient world came from the Etruscans. Their dental practitioners designed and fitted partial dentures which were kept in place with gold wires. These dentures consisted of carved bone and were adornment for the rich in the Italian district of Tuscany as early as 700 B.C. As well as being decorative, they could also be used for chewing and were at least as efficient as dental appliances made many hundreds of years later.

In the time of the Roman Empire, dental replacements by partial dentures were by no means uncommon, partly as a result of the Etruscan techniques being imitated and adapted. Martial (A.D. 40–104) alludes in many of his epigrams to false teeth made of bone and ivory. He admires Laecania, a courtesan, for having teeth as white as snow, but adds they are not natural: "Thais' teeth are black, Laecania's snowy white. How is that? One has her own, the other those you buy." He also refers to wooden teeth and, in addressing a lady, says: "And you lay aside your teeth at night as you do your silken dress."

European Dentistry in Medieval Times

As Europe emerged from the Dark Ages, the influence of the Romans in particular came to the fore. However, little progress in the technique of dentistry or in the general dental hygiene of individuals was made for hundreds of years.

important in the lives of other Presidents too. President Andrew Jackson (1767–1845) had two full sets of teeth made for him, but refused to wear them. He spoke very little in public as a result. In 1877, the Presidential World Cruise was ruined for President Ulysses Grant (1822–85) by the accidental loss overboard of his false teeth. Thereafter he disappointed large crowds by refusing to speak a word to them.

The agonies caused by dental trouble have undoubtedly had a profound effect on the course of history. King Louis XIV of France developed severe mouth infections in later life. He signed the Revocation of the Edict of Nantes, prohibiting Protestants from practicing their religion, while distraught with an abscess. And the well-preserved skeleton in Uppsala Cathedral of King Gustavus Vasa, the sixteenth-century ruler of Sweden, shows huge cavities in the jawbone caused by tooth decay. He made reckless and wild decisions in later life which would have coincided with the agonies caused by his terribly infected mouth.

As early as the seventeenth century, thought was being given to dental hygiene and, by the end of the 1800s, commercially produced preparations were gaining popularity.

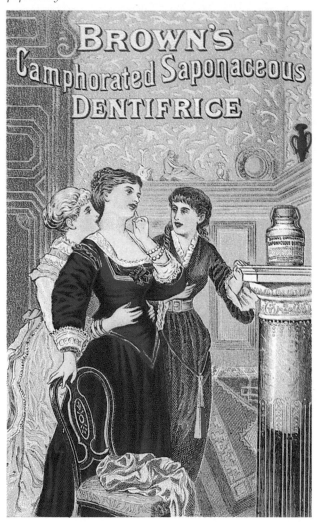

In England, a guild of Barber Surgeons was formed in 1308 which lasted until 1745. These barber surgeons were the main tooth extractors for the population. Some traveled around the countryside wearing necklaces of teeth and promising pain-free extractions; they visited markets and fairs, using pliers (forceps), *pelicans*, and blades, all vicious-looking instruments which no doubt carried infection from one sufferer to the next. The barber surgeon often caused great injury — the jaw was frequently dislocated and sometimes the patient was concussed by an inadvertent blow during extraction. Unstaunchable bleeding and septicemia — blood poisoning — from blunt, rusty and infection ridden instruments were often fatal.

By the time of Shakespeare, the barber surgeons were also taking money for cleaning teeth. They scraped their customers' teeth with metal instruments and then applied a solution of nitric acid called *aqua fortis*. Repeated applications of this simply stripped the enamel surface off the teeth and rendered them highly susceptible to infection. At home, people occasionally used toothpicks or bits of cloth; there were no toothbrushes.

Dentifrice

Around the mid-seventeenth century, the most enlightened surgeons recommended a dentifrice — a contemporary word — made of powdered alabaster and a jelly, presumably applied with the fingers or a cloth. Later, other tooth-cleaning compounds were made up mainly of some abrasive powder such as coral or pumice; these were so gritty that they wore away the enamel as well as the stains which they were supposed to remove. After around 1850, the rich in England carried a quill, or a metal toothpick similar to those used in Roman times.

In France, wooden toothpicks were in use. Jean Renoir, son of the famous painter Auguste Renoir, writes that in the early part of the nineteenth century his grandparents washed out their mouths with salt water, night and morning, and cleaned their teeth with little wooden toothpicks, which were thrown away. (Jean Renoir, *My Father*, 1958)

Transplanting of Teeth

In tooth transplantation, which was practiced at least as early as the sixteenth century, a tooth was removed from one mouth and immediately placed into a vacant socket in another. The poor were often driven to sell their teeth in this way. In Victor Hugo's *Les Miserables*, published in 1862, Fantine's poverty compelled her selling first her hair, then her front teeth and eventually her virtue. Lady Hamilton, mistress to the famous British naval hero, Lord Nelson, was but a young girl when she went one day to sell her teeth. On the way she met a fellow who persuaded her to earn a little money in another well-recognized practice of impoverished girls of the day, and so she was distracted from her original intention.

One macabre practice was the stealing of human teeth from the dead, notably on the battleground.

There are records of nineteenth-century plunderers being paid twenty to thirty pounds — a fortune in those times — for teeth. And many people were provided, unknowingly, with teeth removed from corpses on the field at Waterloo, with infection being transferred with the teeth. In the Civil War, a horde of tooth drawers followed the armies, and the teeth obtained were used in the United States and were also shipped to England. A French surgeon, Pierre le Mayeur, dentist to George Washington, became very wealthy by transplanting teeth for the rich in America, although there is no record of his having transplanted teeth for Washington himself.

Eighteenth-century America

Many dentists left Europe for America at the end of the eighteenth century, but little in the way of restorative dentistry was carried out. Martha Washington, wife of the President, ordered a row of teeth in 1797 for which she had to wait at least two years. The first President himself suffered considerably from his poor teeth, despite the fact that he used sponge toothbrushes. As a young man in his twenties he kept his mouth tightly closed to conceal gaps in his teeth and decayed remains. By his early forties he had partial dentures held in with wire and, within a decade, was virtually toothless. He had a complete set of dentures which were hinged together with strong springs of coiled steel. An English traveler met him in 1790 and wrote: "His mouth was like no other I ever saw; the lips firm and the under jaw seemed to grasp the upper with force, as if the muscles were in full action when he sat still."

Washington was very fond of pickled tripe because it was soft to chew. He sent to London for a supply. "Dental infirmity impels me caring for this necessary item in our domestic commissariat," he wrote. He had several sets of false teeth during

To a large extent, dentistry still is concerned with the repair and maintenance of teeth. When a tooth becomes eroded by plaque produced acid (below), a cavity forms, extending through the tough outer

coat of enamel and often through the dentine and into the pulp. When the dentist detects the cavity, it is first drilled out (below) to remove eroded material. Next, to give the filling some firm anchorage, the

cavity is undercut (below). With modern high speed drills, the actual drilling process is very quick. When the dentist is happy with the shape of the hole, the filling can be inserted. This is made of a material designed

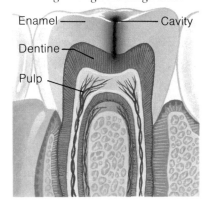

Enamel ——— Cavity
Dentine —
Pulp

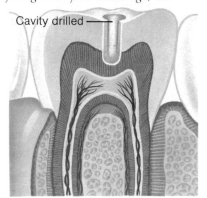

Cavity drilled ———

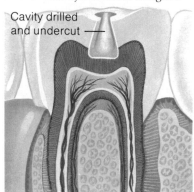

Cavity drilled and undercut ———

his lifetime. For a man acutely sensitive about his appearance and suffering severe toothache for many years, the inadequacies of contemporary dental treatment must have been the cause of a great deal of torment.

The Last Hundred Years

Photographs of our families a hundred years ago shows them to be tight lipped, unsmiling individuals. Many of them are concealing decayed, uneven teeth, for they did not have the enormous benefits of modern developments in dentistry which we now enjoy. Young ladies used to eat in the bedroom before dinner to avoid embarrassment as, often, their false teeth were poor masticators. And not so long before then, false teeth were actually removed at the table before eating. Recommendations concerning the transplantation of human teeth were published as recently as 1919. Today the emphasis of dental care is on prevention of disease; and orthodontics, the straightening of teeth for functional as well as cosmetic reasons, has also rapidly developed.

Dentistry is concerned primarily with two of the world's most common diseases, and both are almost wholly preventable. One, dental caries, attacks the hard enamel coating of the teeth in young and old alike. The other, periodontal disease, attacks the gums and spreads around the tooth sockets causing loosening of the teeth. Both these conditions are primarily caused by plaque.

Plaque is a sticky substance which adheres to the teeth and can be recognized as a "furry feeling" when you run your tongue around your teeth. It is made up of millions of bacteria which thrive on the

debris and food left in the mouth after eating. It accumulates particularly between the teeth and around the gum margins, and it is the toxins from the bacteria which attack and inflame the gingiva (gum), causing gingivitis. The volume of plaque around the teeth and the chemical composition of the deposit are both influenced by what foods are eaten, especially by sugars. Recent research has shown that sucrose is rapidly transformed by plaque bacteria into polysaccharides (starches) which are bulky and expand the volume of the deposit. Small sucrose molecules quickly penetrate the plaque and are broken down by a certain type of acid-forming bacteria. Deep within the plaque, this acid erodes the enamel surface of the tooth, out of reach of the neutralizing alkaline saliva in the mouth. Once the enamel is damaged, the plaque bacteria, notably *Streptococcus mutans*, can penetrate the tooth — this is the start of caries.

Plaque forms more readily on rough surfaces than on smooth, polished ones and will form even in the absence of sugars. However, massive caries-producing plaque only forms in the presence of sugars. And it is not the total quantity of sugar eaten that matters, but the frequency with which it is taken and the time it is in contact with the plaque. Eskimos rarely suffered from tooth decay until they adopted a sugary Western style diet. Moreover, in Europe during World War II, when sugar was scarce, there was little in the way of tooth decay. It is very difficult to alter the habit of a sweet tooth, and for hundreds of years sweet foods have been part of social acts of giving, enjoying and rewarding. Breaking these patterns of behavior requires enormous motivation — sweets and candies are

not to corrode or affect the teeth (below). Finally, the dentist smooths the area over the new filling. A new preventive form of treatment involves the use of sealants. These are especially useful

to prevent decay in children's teeth. First, all teeth are fissured — treated with a special acidic solution to create microscopically fine grooves (below). Then, by applying sealant to this receptive surface, plaque

formation can be inhibited. The sealant (below) bonds to the enamel of the teeth to give tough long-lasting protection. In some instances the sealant may need patching from time to time.

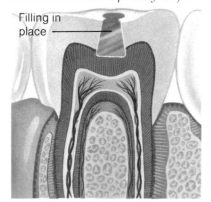

Filling in place

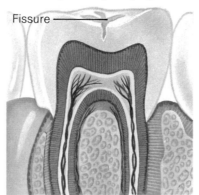

Fissure

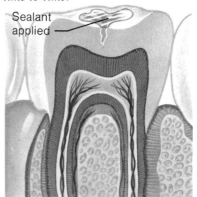

Sealant applied

inexpensive treats for children and sugary carbonated drinks are all too common. A word should be said about fruits. The sugars which give them their sweetness are probably about as dangerous to the teeth as those found in candy, but they do tend to stimulate salivation, which counteracts bacteria produced acids, and the fruits have a slight abrasive cleaning action on the teeth. As a result, they are fairly harmless to the teeth compared with candies and cakes, but an apple at bedtime should never be substituted for a thorough brushing.

Removal of plaque is essential in order to avoid dental caries and gum disease. A number of toothbrushing techniques have been recommended, and all can be effective. It is not so much the technique used as the conscientious application of it that matters in oral hygiene. The only wrong way to brush teeth is the common method employed: horizontal strokes, which wear away the necks of the teeth, damage the gums and jump over the nooks and crannies where the plaque collects. It is essential to try to brush the teeth using short, deliberate, vertical strokes.

Plaque is not easily dislodged so brushing for at least three minutes is the sole effective removal method. The toothbrush itself should not have too large a head, nor too hard bristles; rounded ends to the bristles are desirable. If the brush is rinsed and dried after each use, it may well last two months. When the shape of the brush is lost, it should be discarded — a worn brush is ineffective.

There are other aids, including disclosing tablets and solutions, which reveal where the plaque is on the teeth effectively and dramatically: some even show old and new plaque in different colors. Dental

floss and toothpicks are helpful when used correctly, but overzealous use can damage the gums and cause more harm than good.

These methods reduce the damage done to teeth and gums by the bacterial degradation of sugars. Reduction of sugar intake is important too, but this may be difficult to control in the children and young adults of a relatively wealthy population, and, inevitably, it is these people who suffer most from caries. Modern research, however, is concentrating on other effective measures, such as enzymes to dissipate the plaque, chemotherapy with antibiotics and antiseptics to destroy the acid-forming bacteria, and immunization to build up antibodies to attack these bacteria. A vaccine against plaque bacteria has recently been developed and has been shown to reduce caries by seventy percent in animals. Clinical trials in the U.S.A. and U.K. are currently being carried out, and this form of therapy may provide a much needed breakthrough in the prevention of decay. In the meantime, as we wait for these measures to become generally available, efforts are being made to strengthen tooth enamel.

Fluoride

The enamel of teeth can be toughened by the presence of the element fluoride both during the development of the teeth and after they have erupted. The most effective method of giving fluoride is to include it in the public water supply. Fluoride exists naturally in the water in many regions, and it was noticed that in such areas the teeth of lifelong residents were extremely resistant to decay. Some towns in America have artificially fluoridated

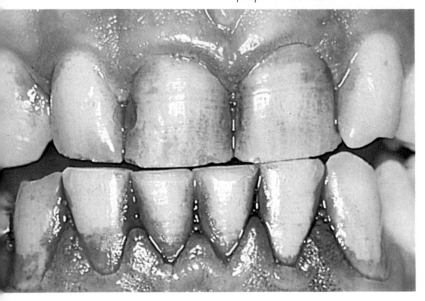

their water for more than thirty years with beneficial effect. No harm has ever been demonstrated as a result of drinking water fluoridated at the level of one part fluoride to every million parts of water (1 ppm; 1 mg/liter).

It is possible and desirable to give children fluoride in the form of tablets or drops from birth to around the age of thirteen, the dose being adjusted according to the amount of fluoride in the local water supply — too much can cause white or brown marks to appear on the teeth. Fluoride toothpaste, mouth rinses and fluoride painted by a dentist directly onto a patient's teeth are all helpful. Another way of aiding nature is by the use of fissure sealants. These have been in use since 1977, and the National Institute of Health is encouraging their use. By filling the pits and fissures of the molar teeth — where plaque thrives — with a hard shield, it should theoretically be feasible to make the development of caries impossible. Two different types of shield material are available, both based on a resin. In both, the enamel is first conditioned by the application of a special buffered phosphoric acid solution. This dissolves certain of the minerals in the surface of the enamel to a depth of about seven microns (roughly one-third of a thousandth of an inch) and leaves a microscopically pitted surface. When the liquid is painted onto this surface, providing it is really clean and dry, the liquid penetrates the minute pits and is polymerized into them. This forms a bond that is durable and impenetrable. The two types of resin differ in their hardening processes; one requires ultraviolet light to be shone on it, the other requires another substance incorporated to form a chemical bond.

It seems that the enamel adhesion either fails in a short time or remains effective for many months or years. It may require patching from time to time and so regular checking is advisable. This apart, almost permanent protection can be given by the fissure sealants.

When it comes to improving the appearance of teeth, rather than preventing decay, there are many new techniques available. One of the most exciting is light cured veneer. Here, discolored or chipped teeth are simply covered with plastic strips and coated with the right shade of a moldable plastic material, which is hardened by a beam of light shone onto it for thirty seconds. This development is replacing the use of crowns in some instances, when much of the tooth has to be cut away and then crowned with porcelain or porcelain bonded with gold. The veneer process is quick, painless and reversible, but the plastic may have to be renewed every few years.

To fill a gap in the teeth, another breakthrough, called the Maryland bridge, has been developed. The dentist roughens the teeth either side of the gap with a bonding agent and then, onto the backs of them, cements two tiny plates which have the replacement tooth suspended between them. In the old style conventional bridge, which is still the preferred treatment for many gaps, the teeth on either side of the gap must be crowned and a false tooth cemented or bonded in the middle. The Maryland bridge causes less damage to adjacent teeth, is quicker to make than a conventional bridge and does not necessitate giving the patient an anesthetic. But it is not always a suitable treatment because it can affect the bite.

Meanwhile, the modern dentist can offer more comfort for those simply seeking routine care. Special jet cleaners are now available, which shoot out bicarbonate of soda in a water jet, instead of the old style scaling and polishing with a brush on the end of a drill. Ultrasound waves are also being used to descale the teeth.

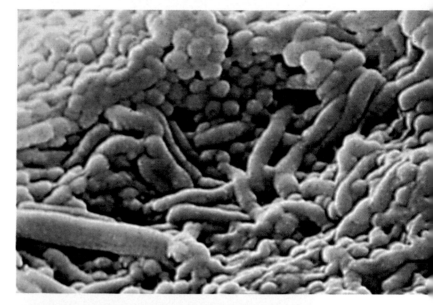

Even that much feared dental implement the drill is being updated. The newest ones have turbine motors which turn the drill at an amazing 350,000 revs a minute and are remarkable in that they are very quiet and do not vibrate.

Dentistry has come a long way since the turn of the century. The current emphasis on prevention rather than cure makes the eradication of the common dental diseases caused by plaque and infection a distinct possibility in the future. Just as smallpox has been eradicated from the world, and tuberculosis is now extremely uncommon in affluent Western societies, so dental caries and gum disease are likely to be phased out by the extraordinary and exciting techniques of present-day science.

Apart from dental disease caused by plaque, problems can arise which are hereditary. We inherit characteristics from each parent, and this can lead to teeth which are either too big or too small for the jaw. When this happens, the services of an orthodontist are required. If the teeth are overcrowded, then it is often necessary to extract a tooth to allow sufficient space for others to grow normally, and irregular teeth can be corrected by using a brace. This type of treatment is more likely to be successful if it is carried out while the teeth and jaws are still developing.

Dietary Fiber

The essence of good dentistry today is a preventive and conservative approach, with every effort being made to halt the progression of caries and plaque, diseases which are in no small way a penalty of the sugar laden diet which we choose to eat. In general, physicians may well be less accustomed to the preventive approach than their dental colleagues, but they too have become increasingly aware of the importance of diet in many of the diseases which they have to treat, both within the alimentary system and outside it. Perhaps the most important breakthrough in recent years is the realization of the importance of dietary fiber. Physicians now prescribe high fiber diets for many conditions, and there is a general move afoot to try to increase fiber in the normal American diet.

Over the last twenty-five years, the growing knowledge about dietary fiber has completely altered the science of nutrition and medical ideas

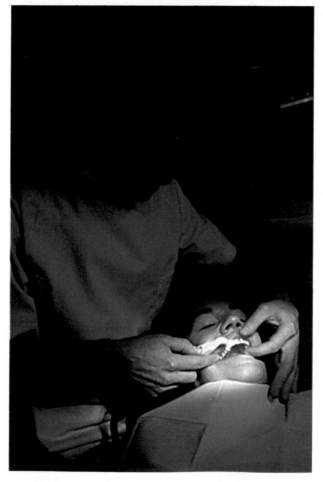

about the causes and treatment of many disorders of the gut. Gradually doctors have come to realize that the diet of the affluent, with its high proportion of expensive foods, such as meat, and highly processed foods to satisfy our so-called sophisticated tastes, rather than doing us any good has actually caused a great deal of harm.

In the past, nutritionists were concerned simply with the nutrients in food. Fiber was ignored or regarded as an unimportant waste material. Nowadays it is the most extensively researched component of our diet and its absence is widely blamed for many of the commonest diseases of modern man.

It was realized first by T.L. Cleave, a Surgeon Commander with the British navy, that there are numerous diseases which are commonplace in Western countries and rare in developing countries. Cleave's hypothesis was that many of these diseases can be explained by the refining of carbohydrate, where foods have their fiber rich fraction taken out and discarded. What remains is a high energy portion that is eaten or drunk. Cleave thought that this processing had two ill effects: that of causing bowel disease by altering the function of the intestines, and of encouraging overnutrition by making high energy foods that can be consumed and digested too easily. There is still a good deal of speculation as to how much this lack of fiber in the diet causes specific diseases. Other factors such as dietary fats and a hereditary tendency are thought by some to be just as important. Despite these uncertainties, a good deal of evidence is emerging regarding those illnesses which are likely to be prevented by a fiber rich diet, and those that can be treated with a high fiber diet.

Fiber in the Gut

Put simply, dietary fiber consists of the undigested remnants of plant cells, particularly the cell walls, which are made of cellulose. These slow down the absorption of nutrients from vegetable material as it passes through the small intestine. When the fiber reaches the colon, however, it is broken down by the bacteria that live in their millions within the large intestine, and during this process carbohydrate is split and gases are released. The best-known constituent of fiber is cellulose, which consists of long strings of glucose units joined together in parallel chains. Some components of fiber are water soluble, for example, the gums and mucilage. Gum is present in oats (resulting in the sticky porridge saucepan) and in some beans.

It is observed that eating fiber increases the weight and volume of the feces, and it was originally thought that the fiber simply caused the feces to retain more water. However, we now know that the situation is more complex. At least half of the fecal weight is made up of bacteria. This proportion increases as the fiber content of the diet increases, since fiber is the main source of nourishment for colonic bacteria.

The amount of fiber necessary to produce a soft stool varies greatly from one individual to another and it is even possible to live a healthy life without fiber, as the Eskimos do. They probably manage to have effortless bowel actions because of the huge amounts of fat in their diet. What is really important, therefore, is to eat as few refined foods as possible — as Cleave originally suggested.

Among the most effective ways of increasing the fiber content of the diet is to eat wholewheat bread. An average slice contains around five grams of

Beans, such as these Mexican beans (below), are not only an excellent and inexpensive source of protein and carbohydrate but they also contain plenty of dietary fiber and no fat whatsoever.

fiber, and six or more slices per day will be required by most people. Some may require even more fiber, which can be added to food as raw bran or by increasing vegetable and fruit intake.

Constipation is a common complaint, and most cases can be helped by a liberal intake of fiber rich foods. If the stool is usually too small or too hard, fear of pain on defecation promotes a vicious circle of constipation, fear, pain and constipation again. Frequency of opening of the bowels is less important than consistency of the stool, but there should probably be at least one bowel movement every other day. Anal fissures and hemorrhoids are frequently caused and often made worse by constipation, which can so easily be corrected.

Irritable bowel syndrome and diverticular disease are two common conditions frequently helped by fiber, especially where constipation is a major factor in the disease. Some people, however, find it difficult to tolerate even a small amount of bran because it can cause a feeling of bloating and discomfort, although this usually disappears after a few weeks of high fiber intake.

One fact noticed recently is that some forms of soluble fiber seem to lower the cholesterol level in the body. Oats, bran and beans taken in large quantities are particularly effective in substantially lowering serum cholesterol. Precisely how this works is unknown, and the importance with regard to prevention and treatment of heart and arterial disease can only be speculated upon.

One has only to visit the urban areas of the United States to observe the condition of obesity in all its variety. There is little scientific evidence to support the view that a reducing diet is more effective if it is rich in fiber, but it is true that fiber rich foods are more satisfying and take longer to digest. Hypoglycemia, where the blood sugar falls very low, is less likely to happen with a regular intake of fiber. And constipation will not be a problem, as it is with a low carbohydrate reducing diet.

Nutritionists now advise us to avoid highly processed fast foods, which are easy to chew, bland, high in fat and salt content, and low in fiber. We must return to a diet containing wholewheat bread, brown rice or some similar, basic carbohydrate food. Such food is more satisfying and more nourishing. It also gives our bodies a better chance of avoiding twentieth-century diseases of the gut and of other body organs.

Appendix

Care of the Teeth

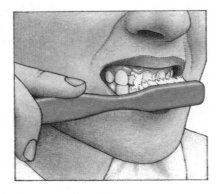
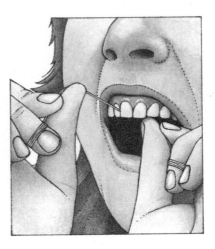

Sweet foods such as these are unhealthy in many ways. As soon as the sugar comes into contact with the teeth, the process of dental decay (caries) can begin. Bacteria cover our teeth with a sticky layer called plaque, which builds up between the teeth and around the gum margins, thriving on food debris in the mouth. In the presence of sugar, the bacteria produce an acid that bores holes into the hard, enamel surface of the teeth.

In the very early stages of dental caries, the saliva can repair small defects in the enamel, but if the teeth are constantly in contact with sugar the saliva does not have an opportunity to repair any damage. All foods containing added sugar are, therefore, dangerous for the teeth, and most decay could be avoided simply by removing refined sugar from the diet. The situation becomes doubly dangerous when sweets are constantly nibbled between meals, since no repair is possible.

In order to preserve healthy teeth, brushing is an essential habit to maintain. Many people brush their teeth in order to refresh their mouths with the taste of dentifrice, rather than for oral hygiene. The object of brushing teeth is to rid the teeth and gum margins of sticky plaque. The plaque accumulates in inaccessible corners and crevices, so the brush must be used on all surfaces of the teeth. Remember that more people lose their teeth from gum disease than from tooth decay, although plaque is still the basic cause. The toothbrush is a valuable aid to the gums; its massaging action prevents gum recession.

Ideally you should brush your teeth after every meal and certainly after the last food or drink of the day. Two or three minutes is the minimum time necessary to remove all the plaque. The toothbrush head should be small, to reach inaccessible crevices, with rounded filaments that will not tear the gums. Try to brush up and down and *between* the teeth to clean out all the plaque.

The prevention of decay and gum diseases is a personal responsibility. If all the plaque can be removed from teeth, you should never suffer any problems.

Dental floss is an invaluable aid to removing debris left after brushing. Floss is a smooth, unbraided thread, which may be either waxed or unwaxed. To use it, cut off a piece about eighteen inches long and wrap it around the middle fingers of each hand, leaving about four to five inches between the hands. Tighten the floss and guide it gently between two teeth, directing it first to one side and then the other, thus cleaning the surface between the two teeth and just below the gum margin.

Clean all the spaces between the teeth in this way. If you press down too hard on the gums they may bleed; however, if they bleed on light pressure, consult your dentist, since this may be a symptom of gum disease. Use dental floss regularly at least twice a week.

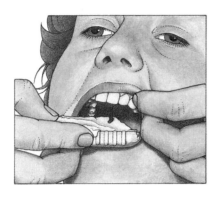  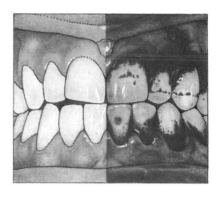

Modern dentistry lays great emphasis on the prevention of disease and there have been many advances in this field. The techniques of repairing teeth cavitated by caries are now greatly improved, with fast and painless techniques based on local anesthetics and the use of high speed drills.

Orthodontic treatment is an important part of the dentist's preventive armament. Its object is to correct and straighten any misalignment of the teeth during childhood and early adolescence. Teeth may be overcrowded and have unsightly gaps, orthodontic correction, however, is more than a simple aesthetic exercise. Overcrowded teeth are more prone to decay, largely because they cannot effectively be cleaned. An unbalanced bite, with uneven patterns of tooth wear, also results from tooth misalignment.

Today it is possible to use almost invisible plastic braces on the teeth, and, on average, it takes one or two years to complete a course of treatment.

Fluoride in dentifrice is an important way of cutting down on tooth caries and on fillings. It has been known for many years that fluoride protects against caries; it exists naturally in the water supply in many localities, and long term residents of these places have demonstrably less dental decay.

Many towns have fluoridated their water over several years with a striking reduction in the incidence of caries. No harm has ever been shown from drinking water with one part per million of flouride, although with more than three parts per million, there may be some yellow or brown staining of the teeth. Children from areas with unfluoridated water should be given supplemental fluorine from birth to around the age of thirteen, with the dose adjusted to the level in the local water supply. Fluoride in dentifrice and mouth washes also helps to give life-long protection against tooth decay.

One of the main problems with plaque is that although it can be felt, it is largely invisible. This is the reason why disclosing tablets and solutions are valuable, particularly in revealing plaque to children.

The disclosing tablet or solution contains a dye which binds to the plaque, making it easily visible. Some of the newer tablets are capable of showing the difference between longstanding plaque and newer plaque, which may have built up in the few hours since the teeth were last brushed. The old plaque is coloured blue by the dye while newer deposits show up red.

Roll the disclosing compound or tablet around the mouth for a short time and rinse out with a small amount of water. Then check in the mirror to see the result. Most people are amazed at the extent of the plaque the first time they try a disclosing agent. After using the agent, brush and floss the teeth in order to clear the colored areas. Disclosing agents are a dramatic way of educating children and adults in care of the teeth.

Eating a healthy diet

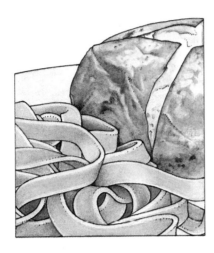
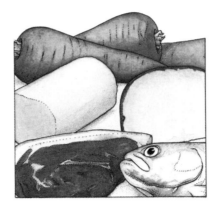

Nutritional science has arrived at the concept of a healthful diet which can be adapted to almost all different environments with their great array of differing food sources.

The basis of the healthful diet is carbohydrate. It is now clear that the human body and its digestive and nutritional systems is designed to run on carbohydrates, but in the last century or so, Western man has moved away from them and has placed a greater reliance on fats and proteins as sources of energy. There is now ample evidence to suggest that this has been partly responsible for the increase in the fatal diseases of the heart and arteries that are so prevalent in Western society, but rare in underdeveloped countries.

In an ideal diet, about half the energy should be derived from carbohydrates, which should be as little refined as possible. Brown rice, wholemeal bread and pasta, potatoes and corn are all excellent carbohydrate sources.

The average American diet is too high in fat from dairy produce and other animal sources, which is disadvantageous for two reasons. First, fat is richer in energy than other food sources, which means that a diet high in fat increases the risk of obesity. Second, it is well known that a high fat diet increases the risk of heart and artery disease, probably because the levels of fat and cholesterol in the blood rise with dietary fat levels. These two problems can be met by a reduction in overall fat intake and a shift toward vegetable oils, fish and poultry as sources of dietary fat.

It is, however, important to retain a balance. A deficiency of fat may well be as dangerous as an excess. Fat is an essential macronutrient and it carries with it important micronutrients such as the fat-soluble vitamins A and D.

The body can manufacture some of the vitamin A it needs, but fish, offal and dairy produce remain the major sources of both vitamins A and D.

The healthfood and vitamin supplementation industry is big business. Thousands of tons of vitamin and mineral pills are consumed every year, and all but a few of these are completely unnecessary.

The key point to appreciate about any dietary vitamin or mineral is that only small amounts are needed each day. An excess over the recommended daily allowance is always unnecessary and may, in fact, be harmful. An ordinary mixed diet will contain all the vitamins and minerals needed for health.

There are four main food groups: the cereals; meat, fish and poultry; dairy produce; and vegetables and fruits. A helping of food from each of these sources every day will provide the full vitamin and mineral support needed.

The two exceptions are iron and fluoride. Iron is necessary in pregnancy, and additional fluoride is needed by children in areas where the fluoride level is low.

Fiber, or roughage as it was once known, has been the Cinderella of nutritional science. Fiber consists of the indigestible elements of foods of plant origin, especially the cellulose-containing plant cell walls.

The role of dietary fiber in health is largely preventive, and fiber seems important to overall health, not just to the digestive tract. The colon needs fiber for normal function, and cholesterol levels and the risk of heart and artery disease are lower with a high fiber diet.

Fiber is most abundant in foods of cereal and vegetable origin and is readily obtainable from all vegetable foods. The difficulty is that over the centuries a habit of food-processing has grown up that aims to remove all possible fiber from fiber rich foods, leaving them less healthful although more digestible. Brown rice, wholemeal bread and peas and beans of all sorts are excellent sources of fiber. Wheat bran, too, is extremely rich in fiber and it may be used as a dietary supplement.

The diet is an element of the twentieth-century lifestyle that has received deserved attention from the health viewpoint. However, it is important to realize that there are other major problems to be grappled with, especially the rather sedentary pursuits of the average town dweller.

Exercise is as important to health as an appropriate diet. A shift toward frequent exercise is also easier to achieve than a wholesale change in eating habits.

The form that exercise takes is significant. The idea is to build on the endurance capacity of the heart, lungs and blood vessels, the so-called "aerobic" system rather than to work on other purely muscular attributes. Activities such as running, swimming and cycling are top of the league in terms of aerobic training. There is no set healthy minimum to the amount of aerobic exercise you need, but three twenty-minute sessions a week should suffice.

People in underdeveloped countries eat little salt in their diet; they also have a low blood pressure which tends to fall with age. In contrast, the inhabitants of the developed world have an extremely high intake of salt in the diet, a higher blood pressure, on average, than most "primitive" peoples and a tendency for blood pressure to rise with increasing age. On their journey from the farm to the table, the most likely thing to happen to most foodstuffs is the addition of salt.

It is easy to add up these facts and, as a result, to label salt as a particular dietary villain and to try and avoid salt altogether. Such a goal would be very hard to achieve and would not necessarily be helpful—the evidence that a population can benefit by cutting its salt intake does not exist.

Extreme views in nutritional areas have nearly always turned out to be wrong. The wise approach is to trim your salt intake to a moderate level of around eight grams a day.

151

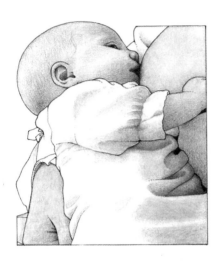

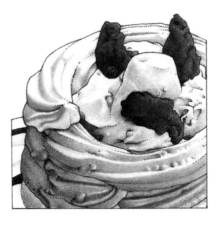

The first few months of human life is the only period in which the body needs anything other than a healthful mixed diet. In the first four months, the diet is one of milk alone. Although formula feeding has improved greatly in the last few decades, breast-feeding is still much to be preferred for a number of reasons.

In the first place, the baby receives valuable antibodies from the mother via the breast milk, and this helps to protect it against infection. Second, a formula cannot precisely mirror the constituents of human milk.

There is also a strong suggestion that the problems of food allergy, which can be very distressing to a few children later in childhood, can be avoided by feeding breast milk alone for the first six months or even longer.

Weaning takes place at different times and rates because babies show varying enthusiasm for additions to their diet.

The early years of childhood are most important in terms of establishing the nutritional habits of a lifetime. Every parent has to reach a decision about the amount of pure sugar in the form of candy, ice cream and cookies that their child is to be allowed. Although children have ample opportunity to frustrate their parents' best intentions, it is nevertheless true that a healthful attitude to food can be developed, despite the wide availability of sweet food to today's children.

The prime reason for avoiding large amounts of sugar is related to dental health; however, there are other important nutritional considerations. If too great a proportion of the daily calorie input is taken as sugar, there will be less of an appetite, and less of a need in energy terms, for the foods that contain important micronutrients (minerals and vitamins) and fiber. Sugary foods are "empty calories," containing only energy (which may be surplus to need) and nothing else of any value.

The hamburgers, french fries and milkshakes—often called "junk foods"—go along with the cakes, cookies and candy as "empty" calorie sources. These foods are not really "junk," they are in many respects perfectly good sources of energy. The problem is that the energy is often surplus to need, so it may lead to obesity and its associated diseases.

Perhaps more important, though, is the fact that these foods contain large amounts of animal fat and salt. Animal fat need not be excluded totally, although there is little doubt that it constitutes too great a proportion of the diet of many Americans, particularly children and teenagers. Similarly there are exceptionally high salt loads in this type of meal that probably contributed to the genesis of hypertension in susceptible individuals.

Junk foods are a real problem in the later stages of childhood and in adolescence, when they become the staple diet of far too many teenagers.

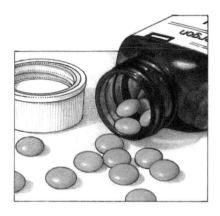

A normal mixed diet contains all the vitamins and minerals that are needed for health. There is no need to supplement vitamin and mineral intake with pills.

Pregnancy is the only major exception to this rule. For women in their reproductive years, there is a constant loss of iron from the body in the course of menstruation. Even on an adequate diet, therefore, the iron levels in the body may be insufficient to meet the huge extra demands that pregnancy makes upon them. Folic acid is a B vitamin which, together with iron, is necessary to maintain normal blood production for both mother and baby. When the supplies of these two micronutrients run low, the developing fetus is kept provided at the expense of the mother, and quite severe anemia can develop in pregnancy. For this reason, combined iron and folic acid pills are part of routine antenatal care. At other times, anemia should be confirmed by medical blood tests before any iron pills are taken.

A diet rich in "junk" foods can cause important nutritional problems in childhood and adolescence, but alcohol can be the major problem throughout adult life. Alcohol is a major drug of dependence, and the risk of addiction and alcoholism are well known.

Alcohol is also a nutritional problem for many people who, quite reasonably, regard themselves as only moderate drinkers.

All alcoholic drinks are high in calories, the alcohol itself provides a lot of energy, and there is often a high sugar content in drinks. In one study, alcohol appeared to provide twenty-five percent of the energy input of a group of nonalcoholic American men. Just like candy and colas these are "empty" calories, and they obviously contribute to the problem of obesity in later life. It is reasonable to keep the level of alcohol down to five percent of the total energy intake: the equivalent of two pints of beer a day.

In later life, nutrition should pose few problems, provided that people continue to eat a healthful mixed diet. The body still needs the same mix of nutrients, although the overall need for energy may be considerably less that it was in the first few decades of life.

As people retire and cease to be as active as they were during their working lives, their energy demands may drop off dramatically. If they do not tailor their food input to meet this reduced demand, they may become increasingly obese. This in turn leads to increasing immobility, thus setting up a vicious circle—immobility further reduces energy needs and inevitably brings about a further increase in weight.

Once this circle is set up, it can be very difficult to break; the answer is to prevent trouble before it occurs. It is probably as important to keep the level of physical activity as high as possible as to cut down on energy input. Any sign of increasing weight, however, should be met by reducing dietary energy.

Glossary

abdomen the space, or cavity, between the chest and the pelvis containing many of the organs of the digestive tract.

achlasia failure of relaxation of the muscle in the esophagus which causes an obstruction to swallowing.

acid a substance that combines with an alkali to form a salt. In the stomach, hydrochloric acid is released and assists in the digestion of food.

adipose describes any body tissue that contains large numbers of fat cells.

aldosterone a hormone produced by the adrenal glands which lie just above the kidneys. It affects the kidneys and plays a role in regulating the fluid balance in the body.

amino acid one of the chemicals from which a protein is built. Twenty different amino acids occur in body proteins in significant quantities.

amylase an enzyme made by the pancreas and small intestine that assists in the digestion of starch.

amyl nitrate a drug administered to reduce the pain of angina. It is taken by inhalation.

amylose a type of carbohydrate; a constituent of starches.

anaerobic surviving or living without oxygen.

anal canal the last section of the digestive tract which opens to the body exterior through the anus.

anemia a condition in which there is a shortage of the oxygen-carrying pigment hemoglobin in the blood. A reduced amount of oxygen thus reaches the tissues.

anesthesia induction of a loss of the power of feeling. It can be general, in which event consciousness is lost, or local, in which sensation is deadened in only part of the body.

antibody a substance found in the blood that is made by the body in response to the presence of a "foreign" substance, such as a toxin or a bacterium. These foreign substances are known as antigens and are destroyed or neutralized by antibodies.

anus the last section of the intestine. It has muscular rings around it to contain feces or flatus. The muscles relax during defecation.

ascites the accumulation of fluid in the abdomen.

bacterium a small microbe. Many bacteria are the agents of disease in the human body.

barium enema/meal barium is a fine, white alkaline earth-metal which is opaque to X rays; barium sulfate can be administered into various sections of the digestive tract, either by mouth or as an enema via the anus. Subsequent X rays will reveal abnormalities in the structure or function of the digestive tract.

bile a thick, brown-green fluid made by the liver, stored in the gallbladder and released to aid digestion of fats.

bilirubin a product, made when hemoglobin is broken down, which helps give bile its color.

biopsy the removal of a piece of body tissue for laboratory examination.

bolus a rounded mass of chewed food that passes down the esophagus and into the stomach when swallowed.

brainstem the bottom section of the brain that joins onto the spinal cord. It controls many body functions below the level of consciousness.

candidiasis a fungal infection often known as "thrush."

carbohydrates a group of substances comprising sugars and starches. The carbohydrates are one of the three types of macronutrients needed by the body.

cardia the part of the stomach nearest to the heart; also the upper opening of the stomach behind the heart.

caries dental decay.

catalyst a substance that speeds up a chemical reaction without itself being changed by that reaction. In the digestive tract, enzymes act as catalysts.

cellulose a carbohydrate that forms the structural element in many plant cells. Dietary fiber is largely composed of this indigestible material. Some animals other than man are able to digest cellulose.

cholecystokinin a hormone produced by the upper intestine. It stimulates the gallbladder to contract and release bile and the pancreas to secrete digestive enzymes.

cholesterol a fatlike substance essential to the continued structure and function of body cell walls. It is generally believed that an excess of cholesterol in the blood can increase the risk of artery disease.

chromosome the DNA-containing part of every cell. On it are borne the genes, which pass on hereditary characteristics.

chyle a milky fluid in which fat is contained. It is found in the lymph vessels of the small intestine.

chyme a grey, acid fluid of homogenized food produced as a result of digestion in the stomach; it passes into the intestine for further digestion.

chymotrypsin a protein-splitting enzyme produced in the pancreas.

cirrhosis a liver disorder in which normal liver tissue becomes replaced by fiber.

colic a pain felt in the abdomen, usually as a result of severe contraction of intestinal muscles.

collateral vessel a subsidiary blood vessel through which blood flows when its normal passage becomes blocked.

colon the part of the large intestine between the cecum and the rectum.

colonoscope a fiberoptic instrument used to see inside the colon.

colostomy the name given to the operation that creates an opening between the colon and the outside of the body, and to the opening itself.

connective tissue the tissue in the body that surrounds, supports, separates and protects the various body organs and structures.

coronary relating to the heart. In a coronary thrombosis a coronary artery becomes blocked, causing a heart attack.

corticosteroid one of a series of hormones produced by the adrenal glands which includes cortisone (cortisone) and aldosterone.

cortisone a hormone responsible for the promotion of glucose production and use, regulation of fat metabolism and the suppression of inflammation.

cryoprobe an instrument that can apply intense cold to the body; used to "freeze off" unwanted tissue.

cyst an abnormal swelling containing fluid.

dentine calcium-containing tissue that surrounds the pulp cavity of a tooth. It comprises the bulk of the tooth.

diaphragm a sheet of muscle dividing the chest cavity from the abdominal cavity.

disaccharide any class of sugars consisting of two linked monosaccharide units per molecule. Sucrose and maltose are examples.

diverticulitis a condition in which the **diverticulae** — small pouches that form in the wall of the colon — become inflamed.

DNA (deoxyribonucleic acid) the basic genetic material that is passed from generation to generation in the genes. DNA is found in cell nuclei and controls protein manufacture in all cells.

duodenum the first part of the small intestine between the stomach and the jejunum.

endocrine system   the system of hormone-producing glands and their hormones. It is involved in controlling many body functions.

enzyme   a protein that promotes a chemical reaction within the body. In digestion, enzymes act as catalysts in the breakdown of large food molecules into small ones capable of being absorbed.

epidemiology   the branch of science concerned with the causes, control and frequency of diseases as they affect populations rather than individuals.

epiglottis   a thin flap of cartilage which covers the entrance to the larynx during swallowing and prevents food from entering the trachea (windpipe).

esophagus   (gullet) the part of the alimentary tract between the pharynx and the stomach.

estrogen   the hormone secreted mainly by the ovaries and the placenta that stimulates changes in the female reproductive organs.

fatty acid   one of the basic units, consisting of long strings of carbon and hydrogen atoms, from which a lipid is constructed.

feces   the end product of the digestive process which contains undigested food, bacteria, water, dead cells and mucus.

fibrosis   a state in which fibrous tissue is formed, as in cirrhosis of the liver.

fistula   any abnormal connection between internal organs or an internal organ and the skin.

flatus   gas, as produced in the intestine.

fluke   a parasitic creature related to the earthworm.

forceps   a surgical instrument with which tissue can be grasped in a pincer grip.

foregut   the section of the growing embryo that will form pharynx, lungs, esophagus, stomach, liver and almost all of the small intestine.

fructose   a simple sugar, easily assimilated and found naturally in fruits.

galactose   a simple sugar sometimes known as milk sugar.

gangrene   death or decay of part of the body most commonly caused by the blocking of the blood supply to the area.

gastrin   a hormone, produced by cells in the lower part of the stomach, that promotes the action of acid-secreting cells.

gastritis   inflammation of the lining of the stomach.

gastroscope   a fiberoptic instrument used to view the inside of the stomach.

gingiva   the soft tissue of the gums. Gingivitis, often caused by poor oral hygiene, is the medical term for inflamed gums.

gland   a body organ producing secretions. An endocrine, or ductless, gland releases its products directly into the blood stream. An exocrine gland usually releases its products via a tube or duct. Exocrine secretions act at sites close to the gland from which they are secreted.

glucose   a simple sugar produced as the end product of starch and carbohydrate digestion. It is the principal source of energy in the body.

glycogen   a carbohydrate stored in the liver that can be converted to glucose to supply energy.

gullet   the esophagus.

gynecomastia   a condition in which the breasts of the male enlarge; it can be a side effect of certain drugs.

halitosis   bad or foul-smelling breath.

hemoglobin   the red, oxygen-carrying pigment in red blood cells.

hemodialysis   the use of a kidney machine to remove waste products from the blood after the patient's own kidneys have stopped functioning.

hemorrhoid   a swollen and twisted vein in the anus or lower rectum.

hepatic artery   the artery that supplies the liver with oxygenated blood.

hepatitis   inflammation of the liver, usually as a result of a virus infection.

hernia   the protrusion of abdominal contents through the wall of the abdominal cavity.

herpes   a group of viruses responsible for causing cold sores, chickenpox, shingles and genital sores.

hiatus hernia   a condition in which the stomach pushes up through the diaphragm into the chest at the point where the gullet is normally located. This is a common cause of chest pain.

hormone   a chemical substance produced in an endocrine gland and carried in the blood stream to a certain tissue, on which it exerts a specific effect.

hydrocarbon   an organic compound which consists of hydrogen and carbon.

ileocecal valve   the valve separating the ileum of the small intestine from the cecum of the large intestine.

ileostomy   operation performed to bypass the large intestine. An opening is made between the patient's skin and the ileum.

ileum   the part of the small intestine between the jejunum and the cecum.

immune system   the body's total defenses against infection.

immunoglobulin   another term for an antibody; a protein produced by the immune defense system that attacks invading organisms.

islets of Langerhans   a group of cells found in the pancreas that is responsible for insulin production.

isoleucine   an amino acid.

jaundice   a yellow skin discoloration caused by bile pigments building up in the blood.

jejunum   the part of the small intestine between the duodenum and the ileum.

Kupffer cells   found in the liver, these cells are largely responsible for the breakdown of hemoglobin into bile pigments.

larynx   (voicebox) situated at the entrance to the windpipe, it contains the vocal cords and is responsible for the production of sound.

leucine   an amino acid.

lingual   of or about the tongue.

lipid   a fat or fatlike substance, generally insoluble in water.

lipoprotein   a substance made up by the combination of a protein and a lipid.

lumen   the space within, or the interior of, a hollow, tube shaped structure of the body.

lymph   a transparent, watery liquid that circulates through the body in a system of tiny vessels in a manner similar to the blood.

lymph nodes   oval or round bodies located in the lymph vessels; they produce lymphocytes.

lymphocyte   one of the types of white blood cell which is found also in the lymph nodes and elsewhere in the body. Lymphocytes are key cells in the body's defense system and are of two types: B lymphocytes are concerned with antibody production, while T lymphocytes are directly involved in the recognition and destruction of invading organisms.

lysine   an amino acid.

M cell   a vital part of the body's immune defense system, these intestinal cells attract toxins and direct them to lymphocytes.

macronutrient   any substance, such as a carbohydrate, which is required in large amounts for healthy growth and development.

mandible   the lower jaw.

membrane   any thin sheet of tissue. Each body cell is surrounded by a wall, or membrane, whose chemical characteristics are central to the way the cell works.

mesentery the double layer of peritoneum which is attached to the back wall of the abdominal cavity. It supports most of the small intestine.

mesoderm the layer of cells in an embryo from which muscle, bone, blood, connective tissue and other structures grow.

methionine an essential amino acid containing sulfur that occurs in many proteins.

micronutrients any substance, such as a vitamin or mineral, which is essential for healthy growth and development but which is required in only small amounts.

microvilli small, fingerlike projections, found on the surface of individual cells in the intestine, which help increase the gut's absorptive surface.

mineral any naturally occuring inorganic substance with a characteristic crystal form. Minerals are essential in the diet, but only in small quantities.

monosaccharide a simple sugar, such as glucose or fructose, which does not react with water to produce other sugars.

occult blood blood — usually passed in the feces — that cannot be seen with the eye. It can be tested for and is usually a sign of disease of the bowel.

orthodontics the correction of crooked or badly positioned teeth.

pancreas an abdominal organ that produces alkaline, enzyme rich fluid, which it discharges into the digestive tract through the pancreatic duct. It also contains the islets of Langerhans.

parathyroid glands endocrine glands whose hormones regulate the metabolism of calcium and phosphorous in the body.

parotid glands two saliva-producing glands; one gland is situated just in front of each ear. They are characteristically swollen during mumps.

pepsin an enzyme produced in the stomach that assists in the digestion of protein.

peptic ulcer an ulcer in the stomach or duodenum.

peristalsis the synchronized contraction of muscles to cause the contents of a tubular organ to be propelled along it. The fundamental mechanism in the intestine that moves food and digestive products along the alimentary canal.

peritoneum the lining membrane that covers the organs contained within the abdominal cavity.

pharynx the throat.

phenylalanine an amino acid.

plaque sticky, yellow-white coating on teeth that contains bacteria. It must be regularly removed by proper dental hygiene if tooth decay is to be avoided.

polyp a growth or tumor — usually benign — arising on a stalk from a mucous membrane. In the digestive system, polyps are found most often in the large intestine.

polysaccharide a substance made up of nine or more simple sugars (monosaccharides).

portal hypertension raised pressure in the vein returning the products of digestion from the intestine to the liver. It is caused by liver disease, especially cirrhosis.

portal vein the vein along which nutrients absorbed from the intestine are transported to the liver for processing. Often known as the hepatic portal vein.

process essentially a prominence or projection. It is a term often used when bones are described.

proctoscope an instrument for viewing the rectum. It usually contains a light and may also have a device for inflating the rectum.

protein a complex organic compound built up of amino acids. Proteins are components of all living matter; a diet containing proteins is essential for the growth and repair of body tissues.

proteolytic describes an enzyme that breaks down proteins into their component parts.

protozoa a group of single celled, usually microscopic animals, some of which are the agents of disease in humans.

purgative a substance that causes evacuation of the bowels.

pylorus the muscular band at the base of the stomach that controls the flow of the stomach contents into the duodenum.

quarantine the isolation of animals or people to prevent the spread of disease.

rectum the lower part of the alimentary canal between the sigmoid colon and the anus.

reflux a flowing back — the reverse flow of acid from the stomach into the esophagus is called esophageal reflux.

rennin an enzyme occurring in gastric juice that clots the protein casein as the first step in the digestion of milk. It is

found in the stomachs of many young mammals, especially ruminants.

RNA (ribonucleic acid) a universal constituent of all living cells which is capable of carrying genetic messages from the chromosomes to the cells.

saturated fat a fat whose fatty acids contain the maximum possible number of hydrogen atoms.

scurvy a disease caused by a lack of vitamin C in the diet.

secretin a hormone secreted by the duodenum when chyme enters the duodenum from the stomach; it stimulates the pancreas to release pancreatic juice.

septicemia (blood poisoning) a disease caused by pathogenic bacteria and their toxins in the blood stream.

sigmoid colon S-shaped section of the colon just before the rectum.

sigmoidoscope an instrument, consisting of a tube with a source of light, which is used to obtain a view of the rectum and the sigmoid colon.

sphincter a circular muscle that controls the opening and closing of a hollow organ.

starch a carbohydrate made up of long chains of glucose molecules. It is found in many plant foods.

stenosis the narrowing of a passage in the body such as a blood vessel or part of the intestine.

steroids a group of hormones having similar, characteristic, chemical structure.

sterol a fatlike steroid, for example cholesterol.

sucrose a simple sugar, a product of the sugar cane and beet plants. The most familiar type of sugar.

suppository medication inserted into body orifices, such as vagina or rectum, in a form that will dissolve easily and be absorbed by the mucosa lining the orifice.

symbiosis a state in which two different organisms live together to their mutual benefit. The bacteria in the human intestine are symbiotic with the person they inhabit. They receive nutrients from the human and, at the same time, manufacture vitamins essential to the human.

threonine an amino acid.

thyroid the gland in the neck whose hormone, thyroxine, is important in controlling the use of energy by the body.

trachea the windpipe.

triglyceride a substance consisting of three fatty-acid molecules linked by a molecule of glycerol. Fats are made up of a series of triglycerides.

trisaccharide a sugar consisting of three monosaccharide molecules linked together.

trypsin an enzyme that occurs in the pancreas and is essential in the digestion of proteins.

tryptophan a sweet-tasting amino acid that occurs in the seeds of some vegetables.

ulcer any break in a smooth lining membrane. Ulcers occur in the intestinal or stomach lining as a result of inflammation and ulceration of the lining of the colon.

unsaturated fat a fat whose constituent fatty acids do not contain the maximum possible number of hydrogen atoms. Extra chemical "space" is taken up by energy rich unsaturated bondings.

urea an end product of the chemical breakdown of proteins in the body. It is excreted in the urine.

vagus nerve a nerve that controls intestinal movement and secretion and thus plays a large part in the digestive process.

valine an amino acid.

varices twisted and enlarged vessels. Esophageal varices are veins in the esophagus that enlarge when a person is suffering from hepatic hypertension.

villi small protrusions from the surface of a membrane with their own blood supply. In the intestine, the villi have an essential role in the absorption of foodstuffs.

virus a microscopic, disease-causing agent that must invade a cell to reproduce itself.

vitamins chemicals occurring naturally in plant and animal tissue and vital in small amounts to the normal function of the body systems.

worms medically, a common term for a large group of parasitic organisms which can survive in the human body. In the digestive system, they include the roundworm *Ascaris* and the threadworm *Enterobius*.

X ray a form of electromagnetic radiation which passes through the body and is absorbed more by some parts — bones for instance — than others. The rays can be recorded on a photographic plate and the resulting pictures enable doctors to observe internal body structures and also give valuable diagnostic information.

yellow fever a serious mosquito spread disease causing jaundice and high fever.

Illustration Credits

Introduction
6, Clark/Goff/Science Photo Library.

The Pathway of Alchemy
8, Ann Ronan Picture Library. 10, (top left) Stammers/Greenwood/Science Photo Library, (top right) Gower Medical Publishing Limited, (bottom) Robert Hessler/Planet Earth Pictures. 11, Michael Holford. 12, Petit Palais, Paris, Bulloz. 13, Bettmann Archive/ BBC Hulton Picture Library. 14, (top) Bettmann Archive/BBC Hulton Picture Library, (bottom) D. B. Lewis/Natural Science Photos. 15, Danish National Museum/Danish Ministry of Foreign Affairs. 16, (left) Carol Hughes/Bruce Coleman Limited, (right) Geoffrey Kinns/Natural Science Photos. 17, Jan and Des Bartlett/Bruce Coleman Limited. 18, (foldout) **Frank Kennard**. 19, (left) Inigo Everson/Bruce Coleman Limited, (right) R. Spoenlein/Zefa UK Limited. 20, Jeff Foott/ Bruce Coleman Limited. 21, Paul Freytag/ Zefa UK Limited. 22, Jan Taylor/Bruce Coleman Limited. 23, Gunter Ziesler/Bruce Coleman Limited. 24, Mark Stanley Price/ Natural Science Photos. 25, (left) **David Gifford**, (right) Photri/Zefa UK Limited. 26–7, **John Bavosi**.

Essential Fuels
28, *An Allegory of the Harvest* by Van Mieris. Johnny van Haeften Gallery/Bridgeman Art Library. 30, (left) *The Last Supper* by Franciabigio. Convent of Santa Maria della Candeli, Florence/Michael Holford, 31, (left) Mark Stanley Price/Natural Science Photos, (right) Forer/Divald/The Image Bank. 32, (top) BBC Hulton Picture Library, (bottom) H. Buchner/Zefa UK Limited. 33, (top) Bettmann Archive/BBC Hulton Picture Library, (bottom) Harald Mante/Zefa UK Limited 34, *The Butcher's Shop* by Annibale Caracci. By courtesy of Christies, London/ Bridgeman Art Library. 35, (top) Nicholas Devore/Bruce Coleman Limited, (bottom) Michael Melford/The Image Bank. 36, (left) José Antonio/The Image Bank, (right) John Topham/Bruce Coleman Limited. 37, (left) David Parker, (right) Peter Newark's Western Americana. 38–9, **Ivan Hissey**. 40, C. Khun/The Image Bank. 41, (left) Frieder Sauer/Bruce Coleman Limited, (right) Marshall Sklar/Science Photo Library. 42, Haslar Naval Hospital, Portsmouth. 43, (left) Martin Dohrn/Science Photo Library, (right) David Parker. 44, **Peter Bridgewater**. 45, Private Collection/Bridgeman Art Library. 46, David Parker. 47, (top left) George Obremski/The Image Bank, (bottom left) Department of Medical Illustration, St. Bartholomews Hospital, London, (right) Zefa UK Limited.

The Journey Begins
48, *The Snack Bar* by Edward Burra/The Tate Gallery, London. 50, (left) *The Five Senses: Taste* by Joe Tilson/The Tate Gallery, London, (right) Gower Medical Publishing Limited. 51, (left) Howard Sochurek/The John Hillelson Agency, (right) Jan Hinsch/ Science Photo Library. 52, **Mike Courtney**. 53, **Mike Courtney**. 54, **Frank Kennard**. 55, BBC Hulton Picture Library. 56, Sygma/The

John Hillelson Agency. 57, Schaefer/Zefa UK Limited. 58, (left) Andrew Edmunds/ MacClancy Collection, (right) James Bell/ Science Photo Library. 59, (left) *The Kiss* by Auguste Rodin. Musée Rodin, Paris/Bulloz, (right) MacClancy Collection. 60–61, **Frank Kennard**. 62, Wellcome Institute Library, London. 63, (top) Ann Ronan Picture Library, (bottom) Gower Medical Publishing Limited. 64, **Mike Courtney**. 65, (top) Dr. J. McFarland/Science Photo Library, (bottom) A. Bode/The Image Bank. 66, (top) Howard Sochurek/The John Hillelson Agency, (bottom) Ann Ronan Picture Library. 67, (all) Smith, Kline & French Laboratories Limited. 68–9, **Frank Kennard**. 71, Wellcome Institute Library, London.

Breaching the Barrier
72, Andrew Edmunds, London/MacClancy Collection. 74, *Claude Bernard and his Pupils* by Léon Lhermitte (copy)/Wellcome Institute Library, London. 75, University College Hospital Library, London. 76, Department of Medical Illustration, St Bartholomews Hospital, London. 77, **Mike Courtney**. 78, Smith, Kline & French Laboratories Limited. 79, **Les Smith**. 80, Department of Medical Illustration, St. Bartholomews Hospital, London. 81, *The Harvest Moon* by Samuel Palmer. By courtesy of Christies, London/Bridgeman Art Library. 82, Wayne Caravella/The Image Bank. 83, Wellcome Institute Library, London, 84, **Mike Courtney**. 85, Wellcome Institute Library, London. 86, Royal Doulton UK Limited. 87, (both) Eric Gravé/Science Photo Library. 88 Gene Cox/Bruce Coleman Limited. 89, Jan Hinsch/Science Photo Library. 91, *The Gates of the Khalif* by William Logsdail. By courtesy of Christies/Bridgeman Art Library.

The Chemical Factory
92, Fotomas Index. 94, (left) British Museum/ Michael Holford, (right) Wellcome Institute Library, London. 95, Archiv für Kunst und Geschichte. 96, **Les Smith**. 97, (left) *Prometheus* by José Ribera. Collection Moussalli/Bulloz, (right) **John Bavosi**. 98, *L'Absinthe* by Edgar Degas. Jeu de Paume, Paris/Bulloz. 99, IGE Medical Systems. 100–101, **John Bavosi**. 102, Bettmann Archive/BBC Hulton Picture Library. 103, Marshall Sklar/Science Photo Library. 104, E. H. Cook/Science Photo Library. 105, (left) E. H. Cook/Science Photo Library, (right) By courtesy of Professor Starzl. 106, **Les Smith**. 107, *The Doctor* by Gerard Dou/Archiv für Kunst und Geschichte. 108, Russ Kinne/ Science Photo Library. 109, Howard Sochurek/The John Hillelson Agency. 110, Gene Cox/Science Photo Library. 111, Martin Dohrn/Science Photo Library.

The Journey Ends
112, Archiv für Kunst und Geschichte. 114, **John Bavosi**. 115, (top) Mansell Collection, (bottom) **John Bavosi**. 117, American Cancer Society. 118, **Les Smith**. 120, **Mike Courtney**. 121, **Mike Courtney**. 122, National Library of Medicine. 123, Department of Medical Illustration, St. Bartholomews Hos-

pital, London. 124, **Frank Kennard**. 125, Department of Medical Illustration, St. Bartholomews Hospital, London. 126, Ann Ronan Picture Library. 127, *La Chasse* by Carle van Loo. By courtesy of Christies, London/Bridgeman Art Library. 128, Archiv für Kunst und Geschichte. 129, (top) Peter Newark's Western Americana, (bottom) Ann Ronan Picture Library.

Maintaining the System
130, Ann Ronan Picture Library. 132, (top) Fotomas Index, (bottom) Mansell Collection. 133, (left) Bulloz, (right) Eric Gravé/Science Photo Library. 134, (left) Ann Ronan Picture Library, (right) Maroon/Zefa UK Limited. 135, Michael Abbey/Science Photo Library. 136, W. von dem Bussche/The Image Bank. 137, (top) Kim Taylor/Bruce Coleman Limited, (bottom) Luis Castañeda/The Image Bank. 138, (left) Louvre, Paris/Bridgeman Art Library, (right) *George Washington* by Gilbert Stuart. National Portrait Gallery, Smithsonian Institution, Washington D.C.; Boston Museum of Fine Arts. 139, Ann Ronan Picture Library. 140, Ann Ronan Picture Library. 141, Fotomas Index. 142–3, **Ivan Hissey**. 144, Sue Lloyd. 145, (top) Gibbs Dental Division. (bottom) Mel Digiacomo/ The Image Bank. 146, Whitney Lane/The Image Bank. 147, Harald Sund/The Image Bank.

Appendix
148–152, **Andrew Popkiewicz**.

Index

159

161